Introduction to
ABSTRACT HARMONIC ANALYSIS

LYNN H. LOOMIS
Harvard University

DOVER PUBLICATIONS, INC.
Mineola, New York

Bibliographical Note

This Dover edition, first published in 2011, is an unabridged republication of the work originally published in 1953 by D. Van Nostrand Company, Inc., New York.

Library of Congress Cataloging-in-Publication Data

Loomis, Lynn H., 1915–
    Introduction to abstract harmonic analysis / Lynn H. Loomis. — Dover ed.
        p. cm.
    Originally published: New York : Van Nostrand, 1953.
    Summary: "Harmonic analysis is a branch of advanced mathematics with applications in such diverse areas as signal processing, medical imaging, and quantum mechanics. This classic monograph is the work of a prominent contributor to the field. Geared toward advanced undergraduates and graduate students, it focuses on methods related to Gelfand's theory of Banach algebra. 1953 edition"— Provided by publisher.
    Includes bibliographical references and index.
    ISBN-13: 978-0-486-48123-4
    ISBN-10: 0-486-4813-9
    1. Algebra, Abstract. I. Title.

QA403.L64 2011
515'.785—dc22

                                                                            2011002271

www.doverpublications.com

PREFACE

This book is an outcome of a course given at Harvard first by G. W. Mackey and later by the author. The original course was modeled on Weil's book [48] and covered essentially the material of that book with modifications. As Gelfand's theory of Banach algebras and its applicability to harmonic analysis on groups became better known, the methods and content of the course inevitably shifted in this direction, and the present volume concerns itself almost exclusively with this point of view. Thus our development of the subject centers around Chapters IV and V, in which the elementary theory of Banach algebras is worked out, and groups are relegated to the supporting role of being the principal application.

A typical result of this shift in emphasis and method is our treatment of the Plancherel theorem. This theorem was first formulated and proved as a theorem on a general locally compact Abelian group by Weil. His proof involved the structure theory of groups and was difficult. Then Krein [29] discovered, apparently without knowledge of Weil's theorem, that the Plancherel theorem was a natural consequence of the application of Gelfand's theory to the L^1 algebra of the group. This led quite naturally to the formulation of the theorem in the setting of an abstract commutative (Banach) algebra with an involution, and we follow Godement [20] in taking this as our basic Plancherel theorem. (An alternate proof of the Plancherel theorem on groups, based on the Krein-Milman theorem, was given by Cartan and Godement [8].)

The choice of Banach algebras as principal theme implies the neglect of certain other tools which are nevertheless important in the investigation of general harmonic analysis, such as the Krein-Milman theorem and von Neumann's direct integral theory, each of which is susceptible of systematic application. It is believed, however, that the elementary theory of Banach algebras and its applications constitute a portion of the subject which in itself is of general interest and usefulness, which can serve as an introduction to present-day research in the whole field, and whose treatment has already acquired an elegance promising some measure of permanence.

The specific prerequisites for the reader include a knowledge of the concepts of elementary modern algebra and of metric space topology. In addition, Liouville's theorem from the elementary theory of analytic functions is used twice. In practice it will probably be found that the requirements go further than this, for without some acquaintance with measure theory, general topology, and Banach space theory the reader may find the preliminary material to be such heavy going as to overshadow the main portions of the book. This preliminary material, in Chapters I–III, is therefore presented in a condensed form under the assumption that the reader will not be unfamiliar with the ideas being discussed. Moreover, the topics treated are limited to those which are necessary for later chapters; no effort has been made to write small textbooks of these subjects. With these restrictions in mind, it is nevertheless hoped that the reader will find these chapters self-contained and sufficient for all later purposes.

As we have mentioned above, Chapter IV is the central chapter of the book, containing an exposition of the elements of the theory of Banach algebras in, it is hoped, a more leisurely and systematic manner than found in the first three chapters. Chapter V treats certain special Banach algebras and introduces some of the notions and theorems of harmonic analysis in their abstract forms. Chapter VI is devoted to proving the existence and uniqueness of the Haar integral on an arbitrary locally compact group, and in Chapters VII and VIII the theory of Banach algebras is applied to deduce the standard theory of harmonic analysis on locally compact Abelian groups and compact groups respectively. Topics covered in Chapter VII include positive definite functions and the generalized Bochner theorem, the Fourier transform and Plancherel theorem, the Wiener Tauberian theorem and the Pontriagin duality theorem. Chapter VIII is concerned with representation theory and the theory of almost periodic functions. We conclude in Chapter IX with a few pages of introduction to the problems and literature of some of the areas in which results are incomplete and interest remains high.

CONTENTS

Chapter I

TOPOLOGY

§ 1. SETS

1A. The reader is assumed to be familiar with the algebra of sets, and the present paragraph is confined to notation. The curly bracket notation is used to name specific sets. Thus $\{a, b\}$ is the set whose elements are a and b, and $\{a_1, \cdots, a_n\}$ is the set whose elements are $a_1, \cdots, a_n$. However, sets cannot generally be named by listing their elements, and more often are defined by properties, the notation being that $\{x: (\ \ )\}$ is the set of all x such that $(\ \ )$. We write "$a \in A$" for "a is a member of A" and "$B \subset A$" for "B is a subset of A"; thus $A = \{x: x \in A\}$.

The ordinary "cup" and "cap" notation will be used for combinations of sets: $A \cup B$ for the union of A and B, and $A \cap B$ for their intersection; $\bigcup_{n=1}^{\infty} A_n$ for the union of the sets of the countable family $\{A_n\}$, and $\bigcap_{n=1}^{\infty} A_n$ for their intersection. More generally, if $\mathfrak{F}$ is any family of sets, then $\bigcup_{A \in \mathfrak{F}} A$ or $\bigcup \{A: A \in \mathfrak{F}\}$ is the union of the sets in $\mathfrak{F}$, and $\bigcap \{A: A \in \mathfrak{F}\}$ is their intersection. The complement of A, designated A', is understood to be taken with respect to the "space" in question, that is, with respect to whatever class is the domain of the given discussion. The null class is denoted by $\varnothing$. The only symbols from logic which will be used are "$\Rightarrow$" for "if $\cdots$ then" and "$\Leftrightarrow$" for "if and only if."

1B. A binary relation "$<$" between elements of a class A is called a *partial ordering* of A (in the weak sense) if it is *transitive* ($a < b$ and $b < c \Rightarrow a < c$), *reflexive* ($a < a$ for every $a \in A$) and if $a < b$ and $b < a \Rightarrow a = b$. A is called a *directed set*

1

under " $<$ " if it is partially ordered by " $<$ " and if for every a and $b \in A$ there exists $c \in A$ such that $c < a$ and $c < b$. More properly, A in this case should be said to be directed *downward;* there is an obvious dual definition for being directed *upward.*

A partially ordered set A is *linearly ordered* if either $a < b$ or $b < a$ for every distinct pair a and $b \in A$. A is partially ordered in the strong sense by " $<$ " if " $<$ " is transitive and *irreflexive* ($a \not< a$). In this case " $\leqq$ " is evidently a corresponding weak partial ordering, and, conversely, every weak partial ordering has an associated strong partial ordering.

1C. The fundamental axiom of set theory which has equivalent formulations in Zorn's lemma, the axiom of choice and the well-ordering hypothesis will be freely used throughout this book. In fact, its use is absolutely essential for the success of certain abstract methods such as found in Gelfand's theory of Banach algebras. Many mathematicians feel dubious about the validity of the axiom of choice, but it must be remembered that Gödel has shown that, if mathematics is consistent without the axiom of choice, then it remains consistent if this axiom is added (and similarly for the generalized continuum hypothesis). Thus a theorem which is proved with the aid of the axiom of choice can never be disproved, unless mathematics already contains an inconsistency which has nothing to do with the axiom of choice.

Zorn's lemma. *Every partially ordered set A includes a maximal linearly ordered subset. If every linearly ordered subset of A has an upper bound in A, then A contains a maximum element.*

The second statement follows from the first upon taking an upper bound x of a maximal linearly ordered subset B. Then x is a maximal element of A, for any properly larger element could be added to B without destroying its linear order, contradicting the maximal character of B. We sketch below a proof of the axiom of choice from Zorn's lemma. Of these two properties, it is a matter of mere preference which is taken as an axiom and which is proved as a theorem. However, the proof given here is much easier than the converse proof.

1D. Theorem (Axiom of choice). *If F is a function with domain D such that $F(x)$ is a non-empty set for every $x \in D$, then* *there exists a function f with domain D such that f(x) $\in$ F(x) for every x $\in$ D.*

Proof. We consider a function to be a class of ordered couples (a graph) in the usual way. Let $\mathfrak{F}$ be the family of functions g such that domain $(g) \subset D$ and such that $g(x) \in F(x)$ for every $x \in$ domain (g). $\mathfrak{F}$ is non-empty, for if x_0 is any fixed element of D and y_0 is any fixed element of $F(x_0)$, then the ordered couple $\langle x_0, y_0 \rangle$ is in $\mathfrak{F}$. (More properly, it is the unit class $\{\langle x_0, y_0 \rangle\}$ which belongs to $\mathfrak{F}$.)

$\mathfrak{F}$ is partially ordered by inclusion, and the union of all the functions (sets of ordered couples) in any linearly ordered sub-family $\mathfrak{F}_0$ of $\mathfrak{F}$ is easily seen to belong to $\mathfrak{F}$ and to be an upper bound of $\mathfrak{F}_0$. Therefore $\mathfrak{F}$ contains a maximal element f. Then domain $(f) = D$, since otherwise we can choose a point $x_0 \in D$ — domain (f) and an element $y_0 \in F(x_0)$, so that $f \cup \{\langle x_0, y_0 \rangle\}$ is a function of $\mathfrak{F}$ which is properly larger than f, a contradiction. This function f, therefore, satisfies the conditions of the theorem.

§ 2. TOPOLOGY

2A. A family $\mathfrak{I}$ of subsets of a space (set) S is called a *topology* for S if and only if:

(a) $\varnothing$ and S are in $\mathfrak{I}$;

(b) if $\mathfrak{I}_1 \subset \mathfrak{I}$, then $\bigcup \{A : A \in \mathfrak{I}_1\} \in \mathfrak{I}$; that is, the union of the sets of any subfamily of $\mathfrak{I}$ is a member of $\mathfrak{I}$;

(c) the intersection of any finite number of sets of $\mathfrak{I}$ is a set of $\mathfrak{I}$.

If $\mathfrak{I}_1$ and $\mathfrak{I}_2$ are two topologies for S, then $\mathfrak{I}_1$ is said to be *weaker* than $\mathfrak{I}_2$ if and only if $\mathfrak{I}_1 \subset \mathfrak{I}_2$.

2B. If $\mathfrak{A}$ is any family of subsets of S, then the topology *generated by* $\mathfrak{A}$, $\mathfrak{I}(\mathfrak{A})$, is the smallest topology for S which includes $\mathfrak{A}$; if $\mathfrak{I} = \mathfrak{I}(\mathfrak{A})$, then $\mathfrak{A}$ is called a *sub-basis* for $\mathfrak{I}$. It follows readily that $A \in \mathfrak{I}(\mathfrak{A})$ if and only if A is $\varnothing$ or S, or if A is a union (perhaps uncountable) of finite intersections of sets in $\mathfrak{A}$. If every set in $\mathfrak{I} = \mathfrak{I}(\mathfrak{A})$ is a union of sets in $\mathfrak{A}$, then $\mathfrak{A}$ is called a *basis* for $\mathfrak{I}$.

2C. If a topology $\mathfrak{I}$ is given for S, then S is called a *topological space* and the sets of $\mathfrak{I}$ are the *open* subsets of S. If A is any subset of S, then the union of all the open subsets of A is called the *interior* of A and is denoted int (A); evidently int (A) is the largest open subset of A, and A is open if and only if $A = $ int (A).

If $p \in$ int (A), then A is said to be a *neighborhood* of p. Neighborhoods are generally, but not always, taken to be open sets. A set of neighborhoods of p is called a *neighborhood basis* for p if every open set which contains p includes a neighborhood of the set.

2D. A subset of S is *closed* (with respect to the topology $\mathfrak{I}$) if its complement is open. It follows that $\varnothing$ and S are closed, that the intersection of any number of closed sets is closed, and that the union of any finite number of closed sets is closed. If A is any subset of S, the intersection of all the closed sets which include A is called the *closure* of A and is commonly denoted $\bar{A}$; evidently $\bar{A}$ is the smallest closed set including A, and A is closed if and only if $A = \bar{A}$. Also, $p \in \bar{A}$ if and only if $p \notin$ int (A'), that is, if and only if every open set which contains p also contains at least one point of A.

2E. If S_0 is a subset of a topological space S, then a topology can be induced in S_0 by taking as open subsets of S_0 the intersection of S_0 with the open subsets of S. This is called the *relative topology* induced in S_0 by the topology of S.

2F. A function f whose domain D and range R are topological spaces is said to be *continuous* at $p_0 \in D$ if $f^{-1}(U) = \{p : f(p) \in U\}$ is a neighborhood of p_0 whenever U is a neighborhood of $f(p_0)$. If f is continuous at each point of its domain, it is said to be *continuous* (on D). It follows that f is continuous if and only if $f^{-1}(U)$ is an open subset of D whenever U is an open subset of R, and also if and only if $f^{-1}(C)$ is closed whenever C is closed.

If f is one-to-one and both f and f^{-1} are continuous, then f is said to be a *homeomorphism* between D and R. Evidently a homeomorphism defines a one-to-one correspondence between the topology $\mathfrak{I}_D$ for D and the topology $\mathfrak{I}_R$ for R.

2G. The following conditions on a subset A of a topological space S are obviously equivalent (by way of complementation):

(a) Every family of open sets which covers A includes a finite subfamily which covers A. (Heine-Borel property.)

(b) Every family of relatively closed subsets of A whose intersection is $\varnothing$ includes a finite subfamily whose intersection is $\varnothing$.

(c) If a family of relatively closed subsets of A has the finite intersection property (that every finite subfamily has non-void intersection), then the family itself has a non-void intersection.

A subset A which has any, and so all, of the above three properties is said to be *compact*. It follows immediately from (b) or (c) that a closed subset of a compact set is compact.

2H. Theorem. *A continuous function with a compact domain has a compact range.*

Proof. If $\{U_\alpha\}$ is an open covering of the range R of f, then $\{f^{-1}(U_\alpha)\}$ is an open covering of the compact domain of f, and hence includes a finite subcovering $\{f^{-1}(U_{\alpha_i})\}$ whose image $\{U_{\alpha_i}\}$ is therefore a finite covering of R. Thus R has the Heine-Borel property and is compact.

2I. An indexed set of points $\{p_\alpha\}$ is said to be *directed* (downward) if the indices form a (downward) directed system. A directed set of points $\{p_\alpha\}$ converges to p if for every neighborhood U of p there exists an index β such that $p_\alpha \in U$ for all $\alpha < \beta$. All the notions of topology can be characterized in terms of convergence. For example, a point p is called *a limit point* of a directed set of points $\{p_\alpha\}$ if for every neighborhood U of p and for every index β there exists $\alpha < \beta$ such that $p_\alpha \in U$. Then it can easily be proved that a space is compact if and only if every directed set of points has at least one limit point.

Little use will be made of these notions in this book and no further discussion will be given.

§3. SEPARATION AXIOMS AND THEOREMS

3A. A Hausdorff space is a topological space in which every two distinct points have disjoint neighborhoods.

Lemma. *If C is a compact subset of a Hausdorff space and $p \notin C$, then there exist disjoint open sets U and V such that $p \in U$ and $C \subset V$.*

Proof. Since S is a Hausdorff space, there exists for every point $q \in C$ a pair of disjoint open sets A_q and B_q such that

$p \in A_q$ and $q \in B_q$. Since C is compact, the open covering $\{B_q\}$ includes a finite subcovering $\{B_{q_i}\}$, and the sets $V = \bigcup B_{q_i}$, $U = \bigcap A_{q_i}$ are as required.

Corollary 1. *A compact subset of a Hausdorff space is closed.*

Proof. By the lemma, if $p \notin C$, then $p \notin \bar{C}$, so that $\bar{C} \subset C$, q.e.d.

Corollary 2. *If $\mathfrak{I}_1$ is a Hausdorff topology for S, $\mathfrak{I}_2$ a compact topology for S and $\mathfrak{I}_1 \subset \mathfrak{I}_2$, then $\mathfrak{I}_1 = \mathfrak{I}_2$.*

Proof. Every $\mathfrak{I}_2$-compact set C is $\mathfrak{I}_1$-compact since every $\mathfrak{I}_1$-covering of C is also a $\mathfrak{I}_2$-covering and so can be reduced. But then C is $\mathfrak{I}_1$-closed since $\mathfrak{I}_1$ is a Hausdorff topology. Thus every $\mathfrak{I}_2$-closed set is $\mathfrak{I}_1$-closed and $\mathfrak{I}_2 = \mathfrak{I}_1$.

3B. A topological space is said to be *normal* if it is a Hausdorff space and if for every pair of disjoint closed sets F_1 and F_2 there exist disjoint open sets U_1 and U_2 such that $F_i \subset U_i$, $i = 1, 2$.

Theorem. *A compact Hausdorff space is normal.*

Proof. The proof is the same as that of the lemma in **3A**. For each $p \in F_1$ there exists, by **3A**, a pair of disjoint open sets U_p, V_p such that $p \in U_p$ and $F_2 \subset V_p$. Since F_1 is compact, the open covering $\{U_p\}$ includes a finite subcovering $\{U_{p_i}\}$ and the sets $U_1 = \bigcup U_{p_i}$, $U_2 = \bigcap V_{p_i}$ are as required.

3C. Urysohn's lemma. *If F_0 and F_1 are disjoint closed sets in a normal space S, then there exists a continuous real-valued function on S such that $f = 0$ on F_0, $f = 1$ on F_1 and $0 \leq f \leq 1$.*

Proof. Let $V_1 = F_1'$. The normality of S implies the existence of an open set $V_{1/2}$ such that $F_0 \subset V_{1/2}$, $\bar{V}_{1/2} \subset V_1$. Again, there exist open sets $V_{1/4}$ and $V_{3/4}$ such that $F_0 \subset V_{1/4}$, $\bar{V}_{1/4} \subset V_{1/2}$, $\bar{V}_{1/2} \subset V_{3/4}$, $\bar{V}_{3/4} \subset V_1$. Continuing this process we define an open set V_r for every proper fraction of the form $m/2^n$, $0 \leq m \leq 2^n$, such that $F_0 \subset V_r$, $\bar{V}_r \subset V_1$ and $\bar{V}_r \subset V_s$ if $r < s$.

We now define the function f as follows: $f(p) = 1$ if p is in none of the sets V_r, and $f(p) = \text{glb}\,\{r: p \in V_r\}$ otherwise. Then

$f = 1$ on F_1, $f = 0$ on F_0, and range $(f) \subset [0, 1]$. If $0 < b \leqq 1$, then $f(p) < b$ if and only if $p \in V_r$ for some $r < b$, so that $\{p : f(p) < b\} = \bigcup_{r < b} V_r$, an open set. Similarly, if $0 \leqq a < 1$, then $f(p) > a$ if and only if $p \notin \bar{V}_r$ for some $r > a$, and $\{p : f(p) > a\} = \bigcup_{r > a} \bar{V}_r{}'$, also an open set. Since the intervals $[0, b)$, $(a, 1]$ and their intersections form a basis for the topology of $[0, 1]$, it follows that the inverse image of every open set is open, i.e., that f is continuous.

3D. A topological space is *locally compact* if every point has a closed compact neighborhood.

A locally compact space S can be made compact by the addition of a single point. In fact, if $S_\infty = S \cup \{p_\infty\}$, where p_∞ is any point not in S, then S_∞ is compact if it is topologized by taking as open sets all the open subsets of S, together with all the sets of the form $O \cup \{p_\infty\}$, where O is an open subset of S whose complement is compact. For if $\{O_\alpha\}$ is an open covering of S_∞, then at least one set O_{α_0} contains p_∞, and its complement is therefore a compact subset C of S. The sets $O_\alpha \cap S$ are open in both topologies and cover C. Therefore, a finite subfamily covers C, and, together with O_{α_0}, this gives a finite subfamily of $\{O_\alpha\}$ covering S_∞. Thus S_∞ has the Heine-Borel property and is compact. It is clear that the original topology for S is its relative topology as a subset of S_∞. S_∞ is called the *one point compactification of S*.

If S is a Hausdorff space, then so is S_∞, for any pair of points distinct from p_∞ is separated by the same pair of neighborhoods as before, while p_∞ is separated from another point p by taking an open set O containing p and having compact closure, so that $\bar{O}'$ is an open subset of S_∞ containing p_∞.

3E. Theorem. *If S is a locally compact Hausdorff space and if C and U are respectively compact and open sets such that $C \subset U$, then there exists a real-valued continuous function on S such that $f = 1$ on C, $f = 0$ on U' and $0 \leqq f \leqq 1$.*

Proof. The proof of **3C** could be modified to yield a proof of **3E**. However, it is sufficient to remark that the one point compactification of S is a Hausdorff space in which C and U' are disjoint compact sets, and **3C** can be directly applied.

§ 4. THE STONE-WEIERSTRASS THEOREM

4A. If S is compact, we shall designate by $\mathcal{C}(S)$ the set of all complex-valued continuous functions on S. If S is locally compact but not compact, $\mathcal{C}(S)$ will designate the set of all complex-valued continuous functions on S which "vanish at infinity," in the sense that $\{p: |f(p)| \geq \epsilon\}$ is compact for every positive ϵ. If S is compactified by adding p_∞ (as in **3D**), then $\mathcal{C}(S)$ becomes the subset of $\mathcal{C}(S_\infty)$ consisting of those functions which vanish at p_∞; hence the phrase "vanishing at infinity." In each of these two cases $\mathcal{C}^R(S)$ will designate the corresponding set of all *real-valued* continuous functions.

4B. An *algebra* A over a field F is a vector space over F which is also a ring and in which the mixed associative law relates scalar multiplication to ring multiplication: $(\lambda x)y = x(\lambda y) = \lambda(xy)$. If multiplication is commutative, then A is called a commutative algebra.

Under the usual pointwise definition of the sum and product of two functions it is evident that $\mathcal{C}(S)$ is a commutative algebra over the complex number field and that $\mathcal{C}^R(S)$ is a commutative algebra over the real number field. If S is compact, the constant functions belong to these algebras, and they each therefore have a multiplicative identity. If S is locally compact but not compact, then $\mathcal{C}(S)$ and $\mathcal{C}^R(S)$ do not have identities. We remark for later use that $\mathcal{C}^R(S)$ is also closed under the lattice operations: $f \cup g = \max(f, g), f \cap g = \min(f, g)$.

4C. Lemma. *Let A be a set of real-valued continuous functions on a compact space S which is closed under the lattice operations $f \cup g$ and $f \cap g$. Then the uniform closure of A contains every continuous function on S which can be approximated at every pair of points by a function of A.*

Proof. Let f be a continuous function which can be so approximated, and, given ϵ, let $f_{p,q}$ be a function in A such that $|f(p) - f_{p,q}(p)| < \epsilon$ and $|f(q) - f_{p,q}(q)| < \epsilon$. Let $U_{p,q} = \{r: f_{p,q}(r) < f(r) + \epsilon\}$ and $V_{p,q} = \{r: f_{p,q}(r) > f(r) - \epsilon\}$. Fixing q and varying p, the open sets $U_{p,q}$ cover S, and therefore a finite subfamily of them covers S. Taking the minimum of the corresponding functions $f_{p,q}$, we obtain a continuous function f_q in A such that $f_q < f + \epsilon$ on S and $f_q > f - \epsilon$ on the open set V_q which is the intersection of the corresponding sets $V_{p,q}$. Now varying q and in a similar manner taking the maximum of a finite number of the functions f_q, we obtain a function f_ϵ in A such that $f - \epsilon < f_\epsilon < f + \epsilon$ on S, q.e.d.

4D. Lemma. *A uniformly closed algebra A of bounded real-valued functions on a set S is also closed under the lattice operations.*

Proof. Since $f \cup g = \max (f, g) = (f + g + |f - g|)/2$, it is sufficient to show that $|f| \in A$ if $f \in A$. We may suppose that $\|f\| = \max_{p \in S} |f(p)| \leq 1$.

The Taylor's series for $(t + \epsilon^2)^{\frac{1}{2}}$ about $t = \frac{1}{2}$ converges uniformly in $0 \leq t \leq 1$. Therefore, setting $t = x^2$, there is a polynomial $P(x^2)$ in x^2 such that $|P(x^2) - (x^2 + \epsilon^2)^{\frac{1}{2}}| < \epsilon$ on $[-1, 1]$. Then $|P(0)| < 2\epsilon$, and $|Q(x^2) - (x^2 + \epsilon^2)^{\frac{1}{2}}| < 3\epsilon$, where $Q = P - P(0)$. But $(x^2 + \epsilon^2)^{\frac{1}{2}} - |x| \leq \epsilon$, so that $|Q(x^2) - |x|| < 4\epsilon$ on $[-1, 1]$. Since Q contains no constant term, $Q(f^2) \in A$ and $\|Q(f^2) - |f|\| < 4\epsilon$. Since A is uniformly closed, $|f| \in A$, q.e.d.

4E. The Stone-Weierstrass Theorem (see [46]). *Let S be a compact space and let A be an algebra of real-valued continuous functions on S which separates points. That is, if $p_1 \neq p_2$, there exists $f \in A$ such that $f(p_1) \neq f(p_2)$. Then the uniform closure $\bar{A}$ of A is either the algebra $\mathcal{C}^R(S)$ of all continuous real-valued functions on S, or else the algebra of all continuous real-valued functions which vanish at a single point p_∞ in S.*

Proof. Suppose first that for every $p \in S$ there exists $f \in A$ such that $f(p) \neq 0$. Then, if $p_1 \neq p_2$, there exists $f \in A$ such that $0 \neq f(p_1) \neq f(p_2) \neq 0$. But then, given any two real numbers a and b, there exists $g \in A$ such that $g(p_1) = a$ and $g(p_2) = b$. (For example, a suitable linear combination of the above f and f^2 will suffice.) Since $\bar{A}$ is closed under the lattice operations (**4D**), it follows from **4C** that $\bar{A}$ contains every continuous real-valued function and $\bar{A} = \mathcal{C}^R(S)$.

There remains the possibility that at some point p_∞ every f in A vanishes. We wish to show that, if g is any continuous function vanishing at p_∞, then $g \in \bar{A}$. But if the constant functions are added to A, the first situation is obtained, and g can be approximated by a function of the form $f + c$, $\| g - (f + c) \| < \epsilon/2$, where $f \in A$ and c is constant. Evaluating at p_∞ we get $| c | < \epsilon/2$, so that $\| f - g \| < \epsilon$. Therefore, $g \in \bar{A}$, q.e.d.

§ 5. CARTESIAN PRODUCTS AND WEAK TOPOLOGY

5A. The Cartesian product $S_1 \times S_2$ of two sets S_1 and S_2 is defined as the set of all ordered pairs $\langle p, q \rangle$ such that $p \in S_1$ and $q \in S_2$,
$$S_1 \times S_2 = \{ \langle p, q \rangle : p \in S_1 \text{ and } q \in S_2 \}.$$

Thus the Cartesian plane of analytical geometry is the Cartesian product of the real line by itself. The definition obviously extends to any finite number of factors: $S_1 \times S_2 \times \cdots \times S_n$ is the set of ordered n-ads $\langle p_1, \cdots, p_n \rangle$ such that $p_i \in S_i$ for $i = 1$, $\cdots$, n. In order to see how to extend the definition further we reformulate the definition just given. We have an index set, the integers from 1 to n, and, for every index i, a space S_i. The ordered n-ad $\langle p_1, \cdots, p_n \rangle$ is simply a function whose domain is the index set, with the restriction that $p_i \in S_i$ for every index i. In general, we consider a non-void index set A and, for every index $\alpha \in A$, we suppose given a non-void space S_α. Then the Cartesian product $\prod_{\alpha \in A} S_\alpha$ is defined as the set of all functions p with domain A such that $p(\alpha) = p_\alpha \in S_\alpha$ for every $\alpha \in A$. The axiom of choice (**1D**) asserts that $\prod_\alpha S_\alpha$ is non-empty.

5B. Let $\{ f_\alpha \}$ be a collection of functions whose ranges S_α are topological spaces and which have a common domain S. If S is to be topologized so that all of these functions are continuous, then $f_\alpha^{-1}(U_\alpha)$ must be open for every index α and every open subset U_α of S_α. The topology generated in S by the totality of all such sets as a sub-basis is thus the weakest topology for S in which all the functions f_α are continuous, and is called the weak topology generated in S by the functions $\{ f_\alpha \}$. In defining a sub-basis for S it is sufficient to use, for each α, only sets U_α in a sub-basis for S_α. It is clear that:

If $S_1 \subset S$ and f_α' is the restriction of f_α to S_1, then the weak topology generated in S_1 by the functions $\{ f_\alpha' \}$ is the relativization to S_1 of the weak topology generated in S by the functions $\{ f_\alpha \}$.

5C. The function f_α which maps every point p in a Cartesian product $S = \prod_\alpha S_\alpha$ onto its coordinate p_α in the αth coordinate space is called the projection of S onto S_α. If the spaces S_α are all topological spaces, then we shall certainly require of any topology in S that all the projections be continuous, and S is conventionally given the weakest possible such topology, that is, the weak topology generated by the projections.

If M is any subset of $\prod_\alpha S_\alpha$, then the relative topology induced in M by the above topology in $\prod_\alpha S_\alpha$ is the weakest topology in which the projections f_α, confined to M, are all continuous.

5D. Theorem (Tychonoff). *The Cartesian product of a family of compact spaces is compact.*

Proof, after Bourbaki. Let $\mathfrak{F}$ be any family of closed sets in $S = \prod_\alpha S_\alpha$ which has the finite intersection property (that the intersection of every finite subfamily of $\mathfrak{F}$ is non-void). We have to show that the intersection of $\mathfrak{F}$ is non-void.

First we invoke Zorn's lemma to extend $\mathfrak{F}$ to a family $\mathfrak{F}_0$ of (not necessarily closed) subsets of S which is maximal with respect to the finite intersection property. The projections of the sets of $\mathfrak{F}_0$ on the coordinate space S_α form a family $\mathfrak{F}_0^\alpha$ of sets in that space having the finite intersection property, and, since S_α is assumed to be compact, there is a point p_α which is in the closure of every set of $\mathfrak{F}_0^\alpha$. Let p be the point in S whose αth coordinate is p_α for each α. We shall show that p is in the closure of every set of $\mathfrak{F}_0$, and therefore is in every set of $\mathfrak{F}$, which will finish the proof. Accordingly, let U be any open set in S containing p. Then there exists (by the definition of the topology in S) a finite set of indices $\alpha_1, \cdots, \alpha_n$, and open sets $U_{\alpha_i} \subset S_{\alpha_i}$, $i = 1, \cdots, n$, such that

$$p \in \bigcap_1^n f_{\alpha_i}^{-1}(U_{\alpha_i}) \subset U,$$

where f_α is the projection of S onto S_α. This implies in particular that $p_{\alpha_i} \in U_{\alpha_i}$, and hence that U_{α_i} intersects every set in $\mathfrak{F}_0^{\alpha_i}$. But then $f_{\alpha_i}^{-1}(U_{\alpha_i})$ intersects every set of $\mathfrak{F}_0$, and so belongs to $\mathfrak{F}_0$ (since $\mathfrak{F}_0$ is maximal with respect to the finite intersection property). But then $\bigcap_1^n f_{\alpha_i}^{-1}(U_{\alpha_i}) \in \mathfrak{F}_0$, for the same reason, and so $U \in \mathfrak{F}_0$. Thus U intersects every set of $\mathfrak{F}_0$, and since U was an arbitrary open set of S containing p it follows that p is in the closure of every set of $\mathfrak{F}_0$, q.e.d.

5E. *The Cartesian product of a family of Hausdorff spaces is a Hausdorff space.*

Proof. If $p \neq q$, then $p_\alpha \neq q_\alpha$ for at least one coordinate α, and, since S_α is a Hausdorff space, there exist disjoint open sets A_α and B_α in S_α such that $p_\alpha \in A_\alpha$ and $q_\alpha \in B_\alpha$. But then $f_\alpha^{-1}(A_\alpha)$ and $f_\alpha^{-1}(B_\alpha)$ are disjoint open sets in $\prod_\alpha S_\alpha$ containing p and q respectively, q.e.d.

5F. The following lemma will be useful in Chapter VI.

Lemma. *Let $f(p, q)$ be a continuous function on the Cartesian product $S_1 \times S_2$ of two Hausdorff spaces S_1 and S_2, and let C and O be respectively a compact subset of S_1 and an open set in range (f). Then the set $W = \{q : f(p, q) \in O$ for all p in $C\}$ is open in S_2.*

Proof. This proof is a third application of the device used to prove **3A** and **3B**. If $q_0 \in W$ is fixed and $p \in C$, there is an open set $U \times V$ containing $\langle p, q_0 \rangle$ on which $f(p, q) \in O$. But C, being compact, can be covered by a finite family $U_1, \cdots, U_n$ of such sets U, and if $V = \bigcap_i^n V_i$ is the intersection of the corresponding V sets, we have $f(p, q) \in O$ for $q \in V$ and $p \in C$. Thus if $q_0 \in W$, then there exists an open set V such that $q_0 \in V \subset W$, proving that W is open.

5G. Theorem. *If $\mathfrak{F}$ is a family of complex-valued continuous functions vanishing at infinity on a locally compact space S, separating the points of S and not all vanishing at any point of S, then the weak topology induced on S by $\mathfrak{F}$ is identical with the given topology of S.*

Proof. Let S_∞ be the one point compactification of S and $\mathfrak{I}_2$ its compact topology. The functions of $\mathfrak{F}$ extend to $\mathfrak{I}_2$-continuous functions on S_∞ which vanish at p_∞. If $\mathfrak{I}_1$ is the weak topology defined on S_∞ by the family of extended functions, then $\mathfrak{I}_1 \subset \mathfrak{I}_2$ by definition. Also $\mathfrak{I}_1$ is a Hausdorff topology, for the extended functions clearly separate the points of S_∞. Therefore, $\mathfrak{I}_1 = \mathfrak{I}_2$ by **3A**, Corollary 2, and the theorem follows upon relativizing to S.

Chapter II

BANACH SPACES

§ 6. NORMED LINEAR SPACES

6A. A *normed linear space* is a vector space over the real numbers or over the complex numbers on which is defined a non-negative real-valued function called the *norm* (the norm of x being designated $\| x \|$) such that

$$\| x \| = 0 \Leftrightarrow x = 0$$

$$\| x + y \| \leq \| x \| + \| y \| \quad \text{(triangle inequality)}$$

$$\| \lambda x \| = | \lambda | \cdot \| x \| \quad \text{(homogeneity)}.$$

A normed linear space is generally understood to be over the complex number field, the real case being explicitly labeled as a real normed linear space. A normed linear space becomes a metric space if the distance $\rho(x, y)$ is defined as $\| x - y \|$, and it is called a *Banach space* if it is complete in this metric, i.e., if whenever $\| x_n - x_m \| \to 0$ as n, $m \to \infty$, then there exists an element x such that $\| x_n - x \| \to 0$ as $n \to \infty$.

The reader is reminded that the (metric space) topology in question has for a basis the family of all open spheres, where the sphere about x_0 of radius r, $S(x_0, r)$, is the set $\{x : \| x - x_0 \| < r\}$. The open spheres about x_0 form a neighborhood basis for x_0.

It follows directly from the triangle inequality that $| \| x \| - \| y \| | \leq \| x - y \|$, so that $\| x \|$ is a continuous function of x.

Euclidean n-dimensional space is a real Banach space with the norm of a point (vector) x taken to be its ordinary length $(\sum x_i^2)^{1/2}$,

13

where $x_1, \cdots, x_n$ are the coordinates (or components) of x. It remains a Banach space if the norm of x is changed to $\|x\|_p = (\sum |x_i|^p)^{1/p}$ for any fixed $p \geqq 1$, though the triangle inequality is harder to prove if $p \neq 2$. These Banach spaces are simple examples of the L^p spaces of measure theory, whose basic theory will be developed completely (though concisely) in Chapter III. We mention now only that L^p is the space of all "integrable" real-valued functions f on a fixed measure space such that the integral $\int |f|^p$ is finite, with $\|f\|_p = \left(\int |f|^p\right)^{1/p}$. Another example of a Banach space is the set of all bounded continuous functions on a topological space S, with $\|f\|_\infty = \mathrm{lub}_{x \in S} |f(x)|$. This norm is called the *uniform* norm because $f_n \to f$ in this norm (i.e., $\|f_n - f\| \to 0$) if and only if the functions f_n converge *uniformly* to f. The fact that this normed linear space is complete is equivalent to the theorem that the limit of a uniformly convergent sequence of bounded continuous functions is itself a bounded continuous function.

6B. The following theorem exhibits one of the most important ways in which new Banach spaces are generated from given ones.

Theorem. *If M is a closed subspace of a normed linear space X, then the quotient vector space X/M becomes a normed linear space if the norm of a coset y is defined as its distance from the origin: $\|y\| = \mathrm{glb}\,\{\|x\| : x \in y\}$. If X is complete, then X/M is complete.*

Proof. First, $\|y\| = 0$ if and only if there exists a sequence $x_n \in y$ such that $\|x_n\| \to 0$. Since y is closed, this will occur if and only if $0 \in y$, so that $\|y\| = 0 \Leftrightarrow y = M$. Next $\|y_1 + y_2\| = \mathrm{glb}\,\{\|x_1 + x_2\| : x_1 \in y_1, x_2 \in y_2\} \leqq \mathrm{glb}\,\{\|x_1\| + \|x_2\|\} = \mathrm{glb}\,\{\|x_1\| : x_1 \in y_1\} + \mathrm{glb}\,\{\|x_2\| : x_2 \in y_2\} = \|y_1\| + \|y_2\|$. Similarly $\|\lambda y\| = |\lambda| \cdot \|y\|$, and X/M is thus a normed linear space.

If $\{y_n\}$ is a Cauchy sequence in X/M, we can suppose, by passing to a subsequence if necessary, that $\|y_{n+1} - y_n\| < 2^{-n}$. Then we can inductively select elements $x_n \in y_n$ such that $\|x_{n+1} - x_n\| < 2^{-n}$, for $\rho(x_n, y_{n+1}) = \rho(y_n, y_{n+1}) < 2^{-n}$. If X is complete, the Cauchy sequence $\{x_n\}$ has a limit x_0, and if

y_0 is the coset containing x_0, then $\| y_n - y_0 \| \leqq \| x_n - x_0 \|$ so that $\{y_n\}$ has the limit y_0. That the original sequence converges to y_0 then follows from the general metric space lemma that, if a Cauchy sequence has a convergent subsequence, then it itself is convergent. Thus X/M is complete if X is complete.

§ 7. BOUNDED LINEAR TRANSFORMATIONS

7A. Theorem. *If T is a linear transformation (mapping) of a normed linear space X into a normed linear space Y, then the following conditions are equivalent:*

1) *T is continuous.*

2) *T is continuous at one point.*

3) *T is bounded. That is, there exists a positive constant C such that $\| T(x) \| \leqq C \| x \|$ for all $x \in X$.*

Proof. If T is continuous at x_0, then there is a positive constant B such that $\| T(x - x_0) \| = \| T(x) - T(x_0) \| \leqq 1$ when $\| x - x_0 \| \leqq B$. Thus $\| T(h) \| \leqq 1$ whenever $\| h \| \leqq B$, and, for any non-zero y, $\| T(y) \| = (\| y \|/B) \| T(y(B/\| y \|)) \| \leqq \| y \|/B$, which is 3) with $C = 1/B$. But then $\| T(x) - T(x_1) \| = \| T(x - x_1) \| \leqq C \| x - x_1 \| < \epsilon$ whenever $\| x - x_1 \| < \epsilon/C$, and T is continuous at every point x_1.

7B. The norm $\| T \|$ of a continuous (bounded) linear transformation is defined as the smallest such bound C. Thus $\| T \| = \text{lub}_{x \neq 0} \| T(x) \|/\| x \|$, and it follows at once that:

The set of all bounded linear transformations of X into Y forms itself a normed linear space. If Y is complete (i.e., a Banach space), then so is this space of mappings.

For example, if $\{T_n\}$ is a Cauchy sequence with respect to the above-defined norm, then $\{T_n(x)\}$ is a Cauchy sequence in Y for every $x \in X$, and, if $T(x)$ is its limit, then it is easy to see that T is a bounded linear transformation and that $\| T_n - T \| \to 0$.

7C. The set $\mathfrak{B}(X)$ of all bounded linear transformations of X into itself is not only a normed linear space but is also an algebra, with the product $T_1 T_2$ defined in the usual way: $T_1 T_2(x) = T_1(T_2(x))$. Moreover, $\| T_1 T_2 \| \leqq \| T_1 \| \cdot \| T_2 \|$, for $\| T_1 T_2 \|$

$= \text{lub} \parallel T_1(T_2(x)) \parallel / \parallel x \parallel \leqq \text{lub} (\parallel T_1 \parallel \cdot \parallel T_2 \parallel \cdot \parallel x \parallel)/ \parallel x \parallel =$
$\parallel T_1 \parallel \cdot \parallel T_2 \parallel$. Any algebra over the complex numbers which has a norm under which it is a normed linear space and in which the above product inequality holds is called a *normed algebra*. A complete normed algebra is called a *Banach algebra*. Thus $\mathfrak{B}(X)$ is a Banach algebra if X is a Banach space.

Another example of a Banach algebra is the space $\mathcal{C}(S)$ of bounded continuous functions on a topological space S, with the uniform norm $\parallel f \parallel_\infty = \text{lub} \{ | f(x) | : x \in S \}$. It is easily checked that $\mathcal{C}(S)$ is a normed algebra, and since a uniformly convergent sequence of continuous functions has a continuous limit, it follows that $\mathcal{C}(S)$ is also complete, and therefore is a Banach algebra.

7D. Theorem. *If N is the nullspace of a bounded linear transformation T, then $\parallel T \parallel$ remains unchanged when T is considered as a transformation with domain X/N.*

Proof. T can be considered to be defined on X/N since, if x_1 and x_2 belong to the same coset y, then $x_1 - x_2 \in N$ and $T(x_1) = T(x_2)$. Let $\parallel\parallel T \parallel\parallel$ be the new norm of T. Then

$$\parallel\parallel T \parallel\parallel = \text{lub}_{y \neq 0} \frac{\parallel T(y) \parallel}{\parallel y \parallel} = \text{lub}_{y \neq 0} \text{lub}_{x \in y} \frac{\parallel T(x) \parallel}{\parallel x \parallel}$$
$$= \text{lub}_{x \notin N} \frac{\parallel T(x) \parallel}{\parallel x \parallel} = \parallel T \parallel.$$

7E. A normed linear space X is said to be the *direct sum* of subspaces M and N if X is algebraically the direct sum of M and N, and if the projections of X onto M and N are both continuous (i.e., the topology in X is the Cartesian product topology of $M \times N$). If X is the direct sum of M and N, then the algebraic isomorphism between M and X/N (which assigns to an element of M the coset of X/N containing it) is also a homeomorphism. For the mapping of M onto X/N is norm-decreasing by the definition of the norm in X/N, and the mapping of X/N onto M has the same norm as the projection on M by **7D**.

7F. We remark for later use that a transformation T which is linear and bounded on a dense subspace M of a normed linear space X, and whose range is a subset of a Banach space, has a unique extension to a bounded linear transformation on the whole of X. This is a special case of the theorem that a uniformly continuous function defined on a dense subset of a metric space has a unique continuous extension to the whole space. An easy direct proof can be given, as follows. If T is defined on the sequence $\{x_n\}$ and $x_n \to x_0$, then the inequality $\| T(x_n) - T(x_m) \| \leqq \| T \| \cdot \| x_n - x_m \|$ shows that the sequence $\{T(x_n)\}$ is Cauchy and hence convergent. If its limit is defined to be $T(x_0)$, it is easy to verify that this definition is unique (and consistent in case T is already defined at x_0) and serves to extend T to the whole of X.

7G. This section is devoted to the important closed graph theorem. Here, for the first time, it is essential that the spaces in question be Banach spaces.

Lemma 1. *Let T be a bounded linear transformation of a Banach space X into a Banach space Y. If the image under T of the unit sphere $S_1 = S(0, 1)$ in X is dense in some sphere $U_r = S(0, r)$ about the origin of Y, then it includes U_r.*

Proof. The set $A = U_r \cap T(S_1)$ is dense in U_r by hypothesis. Let $\bar{y}$ be any point of U_r. Given any $\delta > 0$ and taking $y_0 = 0$, we choose inductively a sequence $y_n \in Y$ such that $y_{n+1} - y_n \in \delta^n A$ and $\| y_{n+1} - \bar{y} \| < \delta^{n+1} r$ for all $n \geqq 0$. There exists, therefore, a sequence $\{x_n\}$ such that $T(x_{n+1}) = y_{n+1} - y_n$ and $\| x_{n+1} \| < \delta^n$. If $\bar{x} = \sum_1^\infty x_n$, then $\| \bar{x} \| < 1/(1 - \delta)$ and $T(\bar{x}) = \sum_1^\infty (y_n - y_{n-1}) = \bar{y}$; that is, the image of the sphere of radius $1/(1 - \delta)$ covers U_r. Thus $U_{r(1-\delta)} \subset T(S_1)$ for every δ, and hence $U_r \subset T(S_1)$, q.e.d.

Lemma 2. *If the image of S_1 under T is dense in no sphere of Y, then the range of T includes no sphere of Y.*

Proof. If $T(S_1)$ is dense in no sphere of Y, then $T(S_n) = \{T(x) : \| x \| < n\} = nT(S_1)$ has the same property. Given any sphere $S \subset Y$, there exists therefore a closed sphere $S(y_1, r_1) \subset S$ disjoint from $T(S_1)$, and then, by induction, a sequence of closed spheres $S(y_n, r_n) \subset S(y_{n-1}, r_{n-1})$ such that $S(y_n, r_n)$ is disjoint from $T(S_n)$. We can also require that $r_n \to 0$, and then the sequence $\{y_n\}$ is Cauchy. Its limit y lies in all the spheres $S(y_n, r_n)$ and hence in none of the sets $T(S_n)$. Since $\bigcup_n T(S_n)$

$= T(X)$, we have proved that $T(X)$ does not include any sphere $S \subset Y$, q.e.d.

Theorem. *If T is a one-to-one bounded linear transformation of a Banach space X onto a Banach space Y, then T^{-1} is bounded.*

Proof. Lemma 2 implies that $T(S_1)$ is dense in some sphere in Y, and therefore, by translation, that $T(S_2)$ is dense in a sphere U_r. But then $U_r \subset T(S_2)$ by Lemma 1, $T^{-1}(U_r) \subset S_2$ and $\| T^{-1} \| \leqq 2/r$, q.e.d.

Corollary. *If T is a linear transformation of a Banach space X into a Banach space Y such that the graph of T is closed (as a subset of the Cartesian product $X \times Y$), then T is bounded.*

Proof. The graph of $T(= \{\langle x, Tx \rangle : x \in X\})$ is by assumption a Banach space under the norm $\| \langle x, Tx \rangle \| = \| x \| + \| Tx \|$. Since the transformation $\langle x, Tx \rangle \to x$ is norm-decreasing and onto X, it follows from the theorem that the inverse transformation $x \to \langle x, Tx \rangle$ is bounded and in particular that T is bounded, q.e.d.

This is the closed graph theorem.

§ 8. LINEAR FUNCTIONALS

8A. If Y is the complex number field (with $\| y \| = | y |$), the space of continuous linear mappings of a normed linear space X into Y is called the *conjugate space* of X, denoted X^*, and the individual mappings are called *linear functionals*. Since the complex number field is complete, it follows from **7B** that

$$X^* \text{ is always a Banach space.}$$

If $F \in X^*$ and N is the nullspace of F, then X/N is one-dimensional, since F becomes one-to-one when transferred to X/N (see **7D**). Otherwise stated, the closed subspace N has deficiency 1 in X.

8B. The well-known Hahn-Banach extension theorem implies the existence of plenty of functionals. We shall use the notation $[A]$ for the linear subspace generated by a subset A of a vector space.

Theorem (Hahn-Banach). *If M is a linear subspace of the normed linear space X, if F is a bounded linear functional on M, and if x_0 is a point of X not in M, then F can be extended to $M + [x_0]$ without changing its norm.*

Proof. We first give the proof when X is a real normed linear space. The problem is to determine a suitable value for $\alpha = F(x_0)$; after that the definition $F(x + \lambda x_0) = F(x) + \lambda \alpha$ for every $x \in M$ and every real λ clearly extends F linearly to $M + [x_0]$. Assuming that $\| F \| = 1$, the requirement on α is that $| F(x) + \lambda \alpha | \leq \| x + \lambda x_0 \|$ for every $x \in M$ and every real $\lambda \neq 0$. After dividing out λ, this inequality can be rewritten

$$-F(x_1) - \| x_1 + x_0 \| \leq \alpha \leq -F(x_2) + \| x_2 + x_0 \|$$

for all $x_1, x_2 \in M$. But $F(x_2) - F(x_1) = F(x_2 - x_1) \leq \| x_2 - x_1 \| \leq \| x_2 + x_0 \| + \| x_1 + x_0 \|$, so that the least upper bound of the left member of the displayed inequality is less than or equal to the greatest lower bound of the right member, and α can be taken as any number in between.

We now deduce the complex case from the real case, following [6]. We first remark that a complex normed linear space becomes a real normed linear space if scalar multiplication is restricted to real numbers, and that the real and imaginary parts, G and H, of a complex linear functional F are each real linear functionals. Also $G(ix) + iH(ix) = F(ix) = iF(x) = -H(x) + iG(x)$, so that $H(x) = -G(ix)$ and $F(x) = G(x) - iG(ix)$. If $\| F \| = 1$ on M, then $\| G \| \leq 1$ on M and, by the above proof, G can be extended to the real linear space $M + [x_0]$ in such a way that $\| G \| \leq 1$. If we similarly add ix_0, we obtain the complex subspace generated by M and x_0 and the real linear functional G defined on it. We now set $F(x) = G(x) - iG(ix)$ on this subspace; we have already observed that this is correct on M. F is obviously a real linear functional on the extended space, and in order to prove that it is complex linear it is sufficient to observe that $F(ix) = G(ix) - iG(-x) = i[G(x) - iG(ix)] = iF(x)$. Finally, if x is given we choose $e^{i\theta}$ so that $e^{i\theta}F(x)$ is real and non-negative, and have $| F(x) | = | F(e^{i\theta}x) | = G(e^{i\theta}x) \leq \| e^{i\theta}x \| = \| x \|$, so that $\| F \| \leq 1$, q.e.d.

8C. It follows from Zorn's lemma that *a functional F such as above can be extended to the whole space X without increasing its norm.* For the extensions of F are partially ordered by inclusion, and the union of any linearly ordered subfamily is clearly an extension which includes all the members of the subfamily and hence is an upper bound of the subfamily. Therefore, there exists a maximal extension by Zorn's lemma. Its domain must be X, since otherwise it could be extended further by **8B**.

In particular, *if x_0 is a non-zero element of X, then there exists a linear functional $F \in X^*$ such that $F(x_0) = \| x_0 \|$ and $\| F \| = 1$.* For the functional $F(\lambda x_0) = \lambda \| x_0 \|$ is defined and of norm one on the one-dimensional subspace generated by x_0, and F can be extended to X as above.

Also, *if M is a closed subspace of X and $x_0 \notin M$, then there exists a functional $F \in X^*$ such that $\| F \| = 1$, $F = 0$ on M and $F(x_0) = d$, where d is the distance from x_0 to M.* For if we define F on $[x_0] + M$ by $F(\lambda x_0 - x) = \lambda d$, then F is linear and $\| F \| = \mathrm{lub}\, | \lambda d | / \| \lambda x_0 - x \| = \mathrm{lub}_{x \in M}\, d / \| x_0 - x \| = d / \mathrm{glb}_{x \in M} \| x_0 - x \| = d/d = 1$. We then extend F as above.

Let $M^\perp$ be the set of all $F \in X^*$ such that $F(x) = 0$ for every $x \in M$. $M^\perp$ is called the *annihilator* of M. The above paragraph implies that *$x \in M$ if and only if $F(x) = 0$ for every $F \in M^\perp$.* This fact might be expressed: $(M^\perp)^\perp = M$.

8D. Theorem. *There is a natural norm-preserving isomorphism (imbedding) $x \to x^{**}$ of a normed linear space into its second conjugate space X^{**} defined by $x^{**}(F) = F(x)$ for every $F \in X^*$.*

Proof. If x is fixed and F varies through X^*, then $x^{**}(F) = F(x)$ is clearly linear on X^*. Since $| x^{**}(F) | = | F(x) | \leqq \| F \| \cdot \| x \|$, we see that x^{**} is bounded, with $\| x^{**} \| \leqq \| x \|$. Since by 8C we can find F such that $F(x) = \| x \|$ and $\| F \| = 1$, it follows that $\| x^{**} \| = \mathrm{lub}_F\, | F(x) | / \| F \| \geqq \| x \|$. Therefore, $\| x^{**} \| = \| x \|$, q.e.d.

Generally X is a proper subspace of X^{**}; if $X = X^{**}$ we say that X is *reflexive.*

8E. Let T be a bounded linear mapping of a normed linear space X into a normed linear space Y. For every $G \in Y^*$, the functional F defined by $F(x) = G(T(x))$ is clearly an element of

X^*. The mapping T^* thus defined from Y^* to X^* is evidently linear; it is said to be the *adjoint* of T. If X is imbedded in X^{**}, then T^{**} is an extension of T, for if $x \in X$ and $G \in Y^*$, then $(T^{**}x^{**})G = x^{**}(T^*G) = (T^*G)x = G(Tx) = (Tx)^{**}G$, showing that $T^{**}(x^{**}) = (Tx)^{**}$.

Since $|F(x)| = |G(T(x))| \leq \|G\| \cdot \|T\| \cdot \|x\|$, we see that $\|T^*G\| = \|F\| \leq \|G\| \cdot \|T\|$, and hence T^* is bounded with $\|T^*\| \leq \|T\|$. Since T^{**} is an extension of T, we have conversely that $\|T\| \leq \|T^{**}\| \leq \|T^*\|$. Therefore, $\|T^*\| = \|T\|$.

8F. We shall need the following theorem in Chapter 4:

Theorem. *If $\{F_n\}$ is a sequence of bounded linear functionals on a Banach space X such that the set of numbers $\{|F_n(x)|\}$ is bounded for every $x \in X$, then the set of norms $\{\|F_n\|\}$ is bounded.*

Proof. It is sufficient to show that the sequence of functionals is bounded in some closed sphere. For if $|F_n(x)| \leq B$ whenever $x \in S(x_0, r) = \{x: \|x - x_0\| < r\}$, then $|F_n(y)| \leq |F_n(y - x_0)| + |F_n(x_0)| \leq 2B$ if $\|y\| < r$, and the norms $\|F_n\|$ have the common bound $2B/r$.

But if $\{F_n\}$ is unbounded in every sphere, we can find inductively a nested sequence of closed spheres $\{S_m\}$ with radii converging to 0 and a subsequence $\{F_{n_m}\}$ such that $|F_{n_m}(x)| > m$ throughout S_m. We do this by first choosing a point P_{m+1} interior to S_m and a functional $F_{n_{m+1}}$ such that $|F_{n_{m+1}}(p_{m+1})| > m + 1$, and then this inequality holds throughout a sphere about p_{m+1} by the continuity of $F_{n_{m+1}}$. Since X is complete, there is a point x_0 in $\bigcap_1^\infty S_m$, and $\{|F_{n_m}(x_0)|\}$ is bounded by hypothesis, contradicting $|F_{n_m}(x_0)| > m$. Therefore, $\{F_n\}$ cannot be unbounded in every sphere, and the proof is finished.

Corollary. *If $\{x_n\}$ is a sequence of elements of a normed linear space X such that $\{|F(x_n)|\}$ is bounded for each $F \in X^*$, then the set of norms $\{\|x_n\|\}$ is bounded.* For X^* is a Banach space and $x_n \in X^{**}$ by **8D**; the above theorem can therefore be applied.

§ 9. THE WEAK TOPOLOGY FOR X^*

9A. The conjugate space of a normed linear space X is a set of complex-valued functions on X and, therefore, is a subset of the Cartesian product space $\prod_{x \in X} C_x$, where the elements $x \in X$ serve as indices and the coordinate space C_x is the complex plane. The relative topology thus induced in X^* is called the weak topology for X^*. We know (see **5C**) that it is the weakest topology for X^* in which the functions x^{**} (the projections of X^* on the coordinate spaces) are all continuous. The sets of the form $\{F \colon |\, F(x) - \lambda_0\,| < \epsilon\}$, depending on x, λ_0, and ϵ, form a subbasis for the weak topology of X^*.

9B. Theorem. *The strongly closed unit sphere in X^* is compact in the weak topology.*

Proof. The strongly closed unit sphere is, of course, the set $S = \{F \colon \|\, F \,\| \leqq 1\}$. The values assumed by these functionals at a point $x \in X$ are complex numbers in the closed circle S_x of radius $\|\, x \,\|$. S is thus a subset of the Cartesian product $\prod_{x \in X} S_x$, which, by the Tychonoff theorem (**5D**), is compact. It is therefore sufficient to show that S is weakly closed in $\prod_{x \in X} S_x$. Accordingly, let G be a function in its closure. Given x, y, and ϵ, the set $\{F \colon |\, F(x) - G(x)\,| < \epsilon\} \cap \{F \colon |\, F(y) - G(y)\,| < \epsilon\} \cap \{F \colon |\, F(x + y) - G(x + y)\,| < \epsilon\}$ is open in $\prod_x S_x$ and contains $G(x)$, and therefore contains an element $F \in S$. But $F(x + y) = F(x) + F(y)$, so that $|\, G(x) + G(y) - G(x + y)\,| < 3\epsilon$. Since ϵ is arbitrary, it follows that G is additive. Similarly it follows that $G(\lambda x) = \lambda G(x)$ and that $|\, G(x)\,| \leqq \|\, x \,\|$. Therefore, $G \in S$, proving that S is weakly closed in $\prod_x S_x$, q.e.d.

9C. There are a number of interesting and important theorems about the weak topology in X^* which will not be needed in this book but which perhaps ought to be sampled in the interests of general education. We confine ourselves to one such situation. It was seen in **8C** that a closed subspace $M \subset X$ is the annihilator of its annihilator ($M = M^{\perp\perp}$). A similar characterization can be given to subspaces of X^* which are closed in the weak topology.

Theorem. *Let M be a subspace of X^*, let $M^\perp$ be the intersection of the nullspaces of the functionals in M and let $(M^\perp)^\perp$ be the set of all functionals in X^* which vanish on $M^\perp$. Then $M = (M^\perp)^\perp$ if and only if M is weakly closed.*

Proof. For every $x \in X$ the functional $x^{**}(F)$ is by definition continuous with respect to the weak topology on X^*, and therefore its nullspace is weakly closed. It follows that, if A is any subset of X, then the set of functionals which vanish on A is an intersection of such nullspaces and hence weakly closed. In particular $(M^\perp)^\perp$ is weakly closed. Since $M \subset (M^\perp)^\perp$, there remains to be shown only that, if F_0 is not in the weak closure of M, then F_0 is not in $(M^\perp)^\perp$, that is, that there exists an $x_0 \in M^\perp$ such that $F_0(x_0) \neq 0$. But if F_0 is not in the weak closure of M, then there exists a weak neighborhood of F_0 not intersecting M, i.e., there exists ϵ and $x_1, \cdots, x_n$ such that no $G \in M$ satisfies $|G(x_i) - F_0(x_i)| < \epsilon$ for every $i = 1, \cdots, n$. The elements x_i map M onto a subspace of complex Euclidean n-space $G \to \langle G(x_1), \cdots, G(x_n) \rangle$, and no point of this subspace is in an ϵ-cube about the point $\langle F_0(x_1) \cdots, F_0(x_n) \rangle$. In particular the subspace does not contain this point and is of dimension at most $n - 1$, so that there exist constants $c_1, \cdots, c_n$ such that $\sum c_i G(x_i) = 0$ for every $G \in M$ and $\sum c_i F_0(x_i) = 1$. Taking $x_0 = \sum_1^n c_i x_i$, these conditions become $G(x_0) = 0$ for every $G \in M$ and $F_0(x_0) = 1$, proving that F_0 is not in $(M^\perp)^\perp$.

Remark: It is possible, by refining this type of argument to show that, if M is such that its intersection with every strongly-closed sphere is weakly closed, then $M = (M^\perp)^\perp$. (See [11].)

§ 10. HILBERT SPACE

10A. A Hilbert space H is a Banach space in which the norm satisfies an extra requirement which permits the introduction of the notion of perpendicularity. Perhaps the neatest formulation of this extra property is the parallelogram law:

$$(1) \qquad \| x + y \|^2 + \| x - y \|^2 = 2[\| x \|^2 + \| y \|^2].$$

A scalar product (x, y) can now be defined,

(2) $4(x, y)$
$$= \| x + y \|^2 - \| x - y \|^2 + i\| x + iy \|^2 - i\| x - iy \|^2,$$

and rather tedious elementary calculations yield the laws:

(3) $(x_1 + x_2, y) = (x_1, y) + (x_2, y);$

(4) $(\lambda x, y) = \lambda(x, y);$

(5) $(x, y) = \overline{(y, x)};$

(6) $(x, x) > 0$ if $x \neq 0.$

The additivity condition (3) can be proved by treating the real and imaginary parts separately, and applying the parallelogram law to four sums of the type $\| x_1 + x_2 + y \|^2 + \| x_1 - x_2 + y \|^2$ to obtain the real part. Repeated applications of (3) lead to $((m/2^n)x, y) = (m/2^n)(x, y)$, and hence, by continuity, to $(ax, y) = a(x, y)$ for positive a. On the other hand, it follows from direct inspection of (2) that $(-x, y) = -(x, y)$ and $(ix, y) = i(x, y)$, completing the proof of (4). The remaining two laws also follow upon direct inspection.

10B. In the applications, however, the scalar product is generally more basic than the norm, and the usual development of the theory of Hilbert space starts off with (3)–(6) as axioms, the parallelogram law now following at once from the definition of $\| x \|$ as $(x, x)^{1/2}$. We also obtain the Schwarz inequality,

(7) $| (x, y) | \leq \| x \| \cdot \| y \|,$

with equality only if x and y are proportional, from the expansion
$0 \leq (x - \lambda y, x - \lambda y) = (x, x) - \lambda(y, x) - \bar{\lambda}(x, y) + | \lambda |^2(y, y),$
by setting $\lambda = (x, x)/(y, x)$. If $(y, x) = 0$, the inequality is trivial. Again, substituting from the Schwarz inequality in the expansion of $\| x + y \|^2 = (x + y, x + y)$, we obtain the triangle inequality:

(8) $\| x + y \| \leq \| x \| + \| y \|,$

with equality only if x and y are proportional, with a non-negative constant of proportionality. Thus H is a normed linear space, and now, rather inelegantly, the final assumption is made that

(9) H is complete in the norm $\| x \| = (x, x)^{1/2}$.

Two elements x and y are said to be orthogonal if $(x, y) = 0$. An immediate consequence of orthogonality is the Pythagorean property: $\| x + y \|^2 = \| x \|^2 + \| y \|^2$.

10C. Theorem. *A closed convex set C contains a unique element of smallest norm.*

Proof. Let $d = \mathrm{glb}\ \{\| x \| : x \in C\}$ and choose $x_n \in C$ such that $\| x_n \| \downarrow d$. Then $(x_n + x_m)/2 \in C$ since C is convex, and so $\| x_n + x_m \| \geqq 2d$.

Since $\| x_n - x_m \|^2 = 2(\| x_n \|^2 + \| x_m \|^2) - \| x_n + x_m \|^2$ by the parallelogram law, it follows that $\| x_n - x_m \| \to 0$ as n, $m \to \infty$. If $x_0 = \lim x_n$, then $\| x_0 \| = \lim \| x_n \| = d$. If x were any other element of C such that $\| x \| = d$, then $(x + x_0)/2$ would be an element of C with norm less than d by (8) above. Therefore, x_0 is unique.

10D. Theorem. *If M is a closed subspace of H, then any element $x \in H$ can be uniquely expressed as a sum $x = x_1 + x_2$ of an element $x_1 \in M$ and an element $x_2 \perp M$. The element x_1 is the best approximation to x by elements of M.*

Proof. If $x \notin M$, the coset $\{x - y : y \in M\}$ might be called the hyperplane through x parallel to M. It is clearly convex and closed, and hence contains a unique element $x - x_1$ closest to the origin (**10C**). Setting $x_2 = x - x_1$, we have $\| x_2 - \lambda y \|^2 \geqq \| x_2 \|^2$ for every $y \in M$ and every complex λ. Setting $\lambda = (x_2, y)/(y, y)$ and expanding, we obtain $-| (y, x_2) | \geqq 0$, and hence $(y, x_2) = 0$. Thus $x_2 \perp M$. This argument can be run backward to deduce the minimum nature of $\| x_2 \|$ from the fact that $x_2 \perp M$, which proves the uniqueness of the decomposition. Of course, this can easily be proved directly.

10E. The set of elements orthogonal to M forms a closed subspace $M^\perp$, and the content of the above theorem is that H is the direct sum of M and $M^\perp$. The element x_1 is called the *projection* of x on M, and the transformation E defined by $E(x) = x_1$

is naturally called the *projection* on M. It is clearly a bounded linear transformation of norm 1 which is idempotent ($E^2 = E$). Its range is M and its nullspace is $M^\perp$.

Lemma. *If A is any linear transformation on H such that, in the above notation, $A(M) \subset M$ and $A(M^\perp) \subset M^\perp$, then $AE = EA$.*

Proof. If $x_1 \in M$, then $Ax_1 \in M$, so that $AEx_1 = Ax_1 = EAx_1$. If $x_2 \in M^\perp$, then $Ax_2 \in M^\perp$, so that $AEx_2 = 0 = EAx_2$. Since any x can be written $x_1 + x_2$, it follows that $AEx = EAx$ for all x, q.e.d.

If M_1 and M_2 are orthogonal closed subspaces, the Pythagorean property $\| x_1 + x_2 \|^2 = \| x_1 \|^2 + \| x_2 \|^2$ for a pair of elements $x_1 \in M_1$ and $x_2 \in M_2$ leads easily to the conclusion that the algebraic sum $M_1 + M_2$ is complete, hence closed. Now let $M_1, \cdots, M_n$ be a finite collection of mutually orthogonal closed subspaces and let M_0 be the orthogonal complement of their (closed) algebraic sum. Then H is the direct sum of the subspaces $M_0, \cdots, M_n$, the component in M_i of a vector x is its orthogonal projection x_i on M_i, and Pythagorean property generalizes to the formula $\| x \|^2 = \| \sum_0^n x_i \|^2 = \sum_0^n \| x_i \|^2$. As a corollary of this remark we obtain Bessel's inequality: *if $M_1, \cdots, M_n$ are orthogonal closed subspaces and x_i is the projection of x on M_i, then $\sum_1^n \| x_i \|^2 \leqq \| x \|^2$.*

10F. Theorem. *Let $\{M_\alpha\}$ be a family, possibly uncountable, of pairwise orthogonal closed subspaces of H, and let M be the closure of their algebraic sum. If x_α is the projection on M_α of an element $x \in H$, then $x_\alpha = 0$ except for a countable set of indices α_n, $\sum_1^\infty x_{\alpha_n}$ is convergent, and its sum is the projection of x on M.*

Proof. Let $\alpha_1, \cdots, \alpha_n$ be any finite set of indices. Since $\sum \| x_{\alpha_i} \|^2 \leqq \| x \|^2$, it follows that $x_\alpha = 0$ except for an at most countable set of indices α_n and that $\sum_1^\infty \| x_{\alpha_i} \|^2 \leqq \| x \|^2$. Setting $y_n = \sum_1^n x_{\alpha_i}$, we have $\| y_n - y_m \|^2 = \sum_m^n \| x_{\alpha_i} \|^2 \to 0$ as $n, m \to \infty$. Hence y_n converges to an element $y \in M$, and, since the projection of y on M_α is clearly x_α, it follows that $x - y$ is orthogonal to M_α for every α and so to M, q.e.d.

10G. Theorem. *For each linear functional $F \in H^*$ there is a unique element $y \in H$ such that $F(x) = (x, y)$.*

Proof. If $F = 0$, we take $y = 0$. Otherwise, let M be the nullspace of F and let z be a non-zero element orthogonal to M. Let the scalar c be determined so that the element $y = cz$ satisfies $F(y) = (y, y)$ (i.e., $c = F(z)/(z, z)$). Since H/M is one-dimensional, every $x \in H$ has a unique representation of the form $x = m + \lambda y$, where $m \in M$. Therefore, $F(x) = F(m + \lambda y) = \lambda F(y) = \lambda(y, y) = (m + \lambda y, y) = (x, y)$, q.e.d.

§ 11. INVOLUTION ON $\mathcal{B}(H)$

11A. We have already observed that the bounded linear transformations of a Banach space into itself form a Banach algebra (**7C**). In a Hilbert space H this normed algebra $\mathcal{B}(H)$ admits another important operation, taking the adjoint (see **8E**). For under the identification of H with H^* (**10G**) the adjoint T^* of a transformation $T \in \mathcal{B}(H)$ is likewise in $\mathcal{B}(H)$. In terms of the scalar product the definition of T^* becomes:

$$(1) \qquad (Tx, y) = (x, T^*y)$$

for all $x, y \in H$. We note the following lemma for later use.

Lemma. *The nullspace of a bounded linear transformation A is the orthogonal complement of the range of its adjoint A^*.* For $Ax = 0$ if and only if $(Ax, y) = 0$ for all $y \in H$, and since $(Ax, y) = (x, A^*y)$, this is clearly equivalent to x being orthogonal to the range of A^*.

11B. Theorem. *The involution operation $T \to T^*$ has the following properties:*
1) $T^{**} = T$
2) $(S + T)^* = S^* + T^*$
3) $(\lambda T)^* = \bar{\lambda} T^*$
4) $(ST)^* = T^*S^*$
5) $\| T^*T \| = \| T \|^2$
6) $(I + T^*T)^{-1} \in \mathcal{B}(H)$, *where I is the identity transformation.*

Proof. Properties 1) to 4) are more or less obvious from (1). For instance 4) follows from $(x, (ST)^*y) = (STx, y) = (Tx, S^*y) = (x, T^*S^*y)$. For 5), we already know that $\| T^*T \| \leq \| T^* \| \cdot \| T \| = \| T \|^2$ (using **8E**), and it remains to be shown that

$\| T^2 \| \leqq \| T^*T \|$.   But $\| T^2 \| = \mathrm{lub}\ (Tx,\ Tx)/\| x \|^2 = \mathrm{lub}$
$(x,\ T^*Tx)/\| x \|^2 \leqq \| T^*T \|$ by Schwarz's inequality.

To prove 6) we notice that

$$\| x \|^2 + \| Tx \|^2 = ((I + T^*T)x, x) \leqq \| (I + T^*T)x \| \cdot \| x \|,$$

so that $\| (I + T^*T)x \| \geqq \| x \|$. In particular $I + T^*T$ is one-to-one, and, since it is clearly self-adjoint, its range is dense in H by **11A**. If y_0 is any element H, we can therefore find a sequence x_n such that $(I + T^*T)x_n \rightarrow y_0$, and since $(I + T^*T)^{-1}$ is norm-decreasing it follows that x_n is a Cauchy sequence and converges to some $x_0 \in H$. Thus $(I + T^*T)x_0 = y_0$, proving that the range of $(I + T^*T)$ is the whole of H. Therefore, $(I + T^*T)^{-1}$ is a bounded linear transformation on H with norm at most one, q.e.d.

11C. If A is any algebra over the complex numbers, a *-operation on A satisfying the above properties (1) to (4) is said to be an *involution*. A Banach algebra with an identity and an involution satisfying all the properties (1) to (7) is said to be a C^*-algebra. Gelfand and Neumark [13] have shown that every C^*-algebra is isometric and isomorphic to an algebra of bounded transformations on a suitable Hilbert space. Thus a subalgebra of $\mathcal{B}(H)$ is the most general C^*-algebra. We shall see in the fifth chapter as an easy corollary of the Gelfand theory that, if a C^*-algebra is commutative, then it is isometric and isomorphic to the algebra of all continuous complex-valued functions on a suitable compact Hausdorff space. An application of this result yields a very elegant proof of the spectral theorem.

11D. The following lemma holds for any pair of linear mappings A and B of a vector space into itself:

Lemma. *If $AB = BA$ and if N and R are the nullspace and range of A respectively, then $B(N) \subset N$ and $B(R) \subset R$.*

Proof. If $x \in N$, then $0 = BAx = A(Bx)$, so that $Bx \in N$. If $x \in R$ then $x = Ay$ for some y, so that $Bx = BAy = ABy \in R$.

Chapter III

INTEGRATION

The theory of Haar measure on locally compact groups is conveniently derived from an elementary integral (linear functional) on the continuous functions which vanish outside of compact sets (see Chapter VI). We shall accordingly begin this chapter by presenting Daniell's general extension of an elementary integral to a Lebesgue integral [9]. The rest of the chapter centers around the L^p spaces and the Fubini theorem.

§ 12. THE DANIELL INTEGRAL

12A. We suppose given a vector space L of bounded real-valued functions on a set S and we assume that L is also closed under the lattice operations $f \cup g = \max (f, g)$ and $f \cap g = \min (f, g)$. We can take absolute values in L since $|f| = f \cup 0 - f \cap 0$. Second, we suppose defined on L a non-negative linear functional I which is continuous under monotone limits. These conditions on I are explicitly as follows:

(1) $$I(f + g) = I(f) + I(g)$$

(2) $$I(cf) = cI(f)$$

(3) $$f \geqq 0 \Rightarrow I(f) \geqq 0$$

(3′) $$f \geqq g \Rightarrow I(f) \geqq I(g)$$

(4) $$f_n \downarrow 0 \Rightarrow I(f_n) \downarrow 0$$

Here "$f_n \downarrow$" means that the sequence f_n is pointwise monotone decreasing, i.e., that $f_{n+1}(p) \leqq f_n(p)$ for all n and p, and "$f_n \downarrow f$" means that f_n decreases pointwise to the limit f.

Such a functional will be called an *integral*, to avoid confusion with the bounded linear functionals of normed linear space theory. Our task is to extend I to a larger class of functions having all the properties of L, and which is in addition closed under certain countable operations.

As an example, L can be taken to be the class of continuous functions on [0, 1] and I to be the ordinary Riemann integral. Properties (1) to (3) are well known, and (4) follows from the fact that on a compact space pointwise monotone convergence implies uniform convergence (see **16A**). The extension of L is then the class of Lebesgue summable functions, and the extended I is the ordinary Lebesgue integral.

12B. Every increasing sequence of real-valued functions is convergent if $+\infty$ is allowed as a possible value of the limit function. Let U be the class of limits of monotone increasing sequences of functions of L. U includes L, since the sequence $f_n = f$ is trivially increasing to f as a limit. It is clear that U is closed under addition, multiplication by non-negative constants, and the lattice operations.

We now want to extend I to U, and we make the obvious definition: $I(f) = \lim I(f_n)$, where $f_n \uparrow f$ and $f_n \in L$, and where $+\infty$ is allowed as a possible value of I. It will be shown in **12C** that this number is independent of the particular sequence $\{f_n\}$ converging to f. Assuming this, it is evident that the new definition agrees with the old in case $f \in L$ (for then we can set $f_n = f$) and that the extended I satisfies (1) and (2) with $c \geqq 0$. It will also follow from **12C** that I satisfies (3').

12C. Lemma. *If* $\{f_n\}$ *and* $\{g_m\}$ *are increasing sequences of functions in L and* $\lim g_m \leqq \lim f_n$, *then* $\lim I(g_m) \leqq \lim I(f_n)$.

Proof. We first notice that, if $k \in L$ and $\lim f_n \geqq k$, then $\lim I(f_n) \geqq I(k)$, for $f_n \geqq f_n \cap k$ and $f_n \cap k \uparrow k$, so that $\lim I(f_n) \geqq \lim I(f_n \cap k) = I(k)$, by (3') and (4).

Taking $k = g_m$ and passing to the limit as $m \to \infty$, it follows that $\lim I(f_n) \geqq \lim I(g_m)$, proving **12C**. If $\lim f_n = \lim g_m$, we also have the reverse inequality, proving that I is uniquely defined on U.

12D. Lemma. *If $f_n \in U$ and $f_n \uparrow f$, then $f \in U$ and $I(f_n)$ $\uparrow I(f)$.*

Proof. Choose $g_n{}^m \in L$ so that $g_n{}^m \uparrow_m f_n$, and set $h_n = g_1{}^n$ $\cup \cdots \cup g_n{}^n$. Then $h_n \in L$ and $\{h_n\}$ is an increasing sequence. Also $g_i{}^n \leqq h_n \leqq f_n$ if $i \leqq n$. If we pass to the limit first with respect to n and then with respect to i, we see that $f \leqq \lim h_n \leqq f$. Thus $h_n \uparrow f$ and $f \in U$. Doing the same with the inequality $I(g_i{}^n) \leqq I(h_n) \leqq I(f_n)$, we get $\lim I(f_i) \leqq I(f) \leqq \lim I(f_n)$, proving that $I(f_n) \uparrow I(f)$, q.e.d.

12E. Let $-U$ be the class of negatives of functions in U: $-U = \{f : -f \in U\}$. If $f \in -U$, we make the obvious definition $I(f) = -I(-f)$. If f is also in U, this definition agrees with the old, for $f + (-f) = 0$ and I is additive on U. The class $-U$ obviously has properties exactly analogous to those of U. Thus, $-U$ is closed under monotone decreasing limits, the lattice operations, addition and multiplication by non-negative constants, and I has on $-U$ the properties (1), (2) and (3′). It is important to remark that, if $g \in -U$, $h \in U$, and $g \leqq h$, then $h - g \in U$ and $I(h) - I(g) = I(h - g) \geqq 0$.

A function f is defined to be *summable* (better, *I-summable*) if and only if for every $\epsilon > 0$ there exist $g \in -U$ and $h \in U$ such that $g \leqq f \leqq h$, $I(g)$ and $I(h)$ are finite, and $I(h) - I(g) < \epsilon$. Varying g and h, it follows that glb $I(h) =$ lub $I(g)$, and $I(f)$ is defined to be this common value. The class of summable functions is designated L^1 (or $L^1(I)$); it is the desired extension of L.

We note immediately that, if $f \in U$ and $I(f) < \infty$, then $f \in L^1$ and the new definition of $I(f)$ agrees with the old. For then there exist $f_n \in L$ such that $f_n \uparrow f$, and we can take $h = f$ and $g = f_n$ for suitably large n.

12F. Theorem. *L^1 and I have all the properties of L and I.*

Proof. Given f_1 and $f_2 \in L^1$ and given $\epsilon > 0$, we choose g_1 and $g_2 \in -U$, and $h_1, h_2 \in U$, such that $g_i \leqq f_i \leqq h_i$ and $I(h_i) - I(g_i) < \epsilon/2$, $i = 1, 2$. Letting o be any of the operations $+$, $\cap$, $\cup$, we have $g_1 \circ g_2 \leqq f_1 \circ f_2 \leqq h_1 \circ h_2$ and $h_1 \circ h_2 - g_1 \circ g_2 \leqq$

$(h_1 - g_1) + (h_2 - g_2)$, and therefore $I(h_1 \circ h_2) - I(g_1 \circ g_2) < \epsilon$. It follows at once that $f_1 + f_2, f_1 \cup f_2$ and $f_1 \cap f_2$ are in L^1, and, since I is additive on U and $-U$, that $|I(f_1 + f_2) - I(f_1) - I(f_2)| < 2\epsilon$. Since ϵ is arbitrary, (1) follows. That $cf \in L^1$ and $I(cf) = cI(f)$ if $f \in L^1$ is clear; we remark only that the roles of the approximating functions are interchanged if $c < 0$. If $f \geq 0$, then $h \geq 0$ and $I(f) = \text{glb } I(h) \geq 0$, proving (3). (4) follows from the more general theorem below.

12G. Theorem. *If $f_n \in L^1$ $(n = 0, 1, \cdots)$, $f_n \uparrow f$ and* $\lim I(f_n) < \infty$, *then $f \in L^1$ and $I(f_n) \uparrow I(f)$.*

Proof. We may suppose, by subtracting off f_0 if necessary, that $f_0 = 0$. We choose $h_n \in U$ $(n = 1, \cdots)$ so that $(f_n - f_{n-1}) \leq h_n$ and $I(h_n) < I(f_n - f_{n-1}) + \epsilon/2^n$. Then $f_n \leq \sum_1^n h_i$ and $\sum_1^n I(h_i) < I(f_n) + \epsilon$. Setting $h = \sum_1^\infty h_i$, we have $h \in U$ and $I(h) = \sum_1^\infty I(h_i)$ by **12D**. Moreover, $f \leq h$ and $I(h) \leq \lim I(f_n) + \epsilon$. Thus if m is taken large enough, we can find $g \in -U$ so that $g \leq f_m \leq f \leq h$ and $I(h) - I(g) < 2\epsilon$. It follows that $f \in L^1$ and that $I(f) = \lim I(f_m)$.

12H. A family of real-valued functions is said to be *monotone* if it is closed under the operations of taking monotone increasing and monotone decreasing limits. The smallest monotone family including L will be designated $\circledR$ and its members will be called *Baire* functions.

If $h \leq k$, then any monotone family $\mathfrak{M}$ which contains $(g \cup h) \cap k$ for every $g \in L$ also contains $(f \cup h) \cap k$ for every $f \in \circledR$, for the functions f such that $(f \cup h) \cap k \in \mathfrak{M}$ form a monotone family which includes L and therefore includes $\circledR$. In particular the smallest monotone family including L^+ is $\circledR^+$ (where, for any class of functions C, C^+ is the class of non-negative functions in C).

Theorem. *$\circledR$ is closed under the algebraic and lattice operations.*

Proof. (From Halmos [23].) For any function $f \in \circledR$ let $\mathfrak{M}(f)$ be the set of functions $g \in \circledR$ such that $f + g, f \cup g$ and $f \cap g \in \circledR$. $\mathfrak{M}(f)$ is clearly a monotone family, and, if $f \in L$, then $\mathfrak{M}(f)$ includes L and therefore is the whole of $\circledR$. But $g \in \mathfrak{M}(f)$ if and only if $f \in \mathfrak{M}(g)$; therefore, $\mathfrak{M}(g)$ includes L

for any $g \in \mathcal{B}$. Therefore, $\mathfrak{M}(g) = \mathcal{B}$ for any $g \in \mathcal{B}$. That is, if f and $g \in \mathcal{B}$, then $f + g, f \cup g$ and $f \cap g \in \mathcal{B}$.

Similarly, if $\mathfrak{M}$ is the class of functions $f \in \mathcal{B}$ such that $cf \in \mathcal{B}$ for every real c, then $\mathfrak{M}$ is monotone and includes L so that $\mathfrak{M} = \mathcal{B}$.

12I. A function f will be said to be *L-bounded* if there exists $g \in L^+$ such that $|f| \leqq g$. A family $\mathfrak{F}$ of functions will be said to be *L-monotone*, if whenever f_n is a sequence of L-bounded functions of $\mathfrak{F}$ and $f_n \uparrow f$ or $f_n \downarrow f$, then $f \in \mathfrak{F}$.

Lemma. *If $f \in \mathcal{B}$, then there exists $g \in U$ such that $f \leqq g$.*

Proof. The family of functions $f \in \mathcal{B}$ for which this is true is monotone (by **12D**) and includes L, and is therefore equal to $\mathcal{B}$.

Theorem. *The smallest L-monotone family including L^+ is $\mathcal{B}^+$.*

Proof. Let $\mathfrak{F}$ be this smallest family. For any fixed $g \in L^+$ the functions $f \in \mathcal{B}^+$ such that $f \cap g \in \mathfrak{F}$ form a monotone family including L^+ and therefore equal to $\mathcal{B}^+$. Thus if $f \in \mathcal{B}^+$ and $f \leqq g$, then $f = f \cap g \in \mathfrak{F}$; that is, $\mathfrak{F}$ contains every L-bounded function of $\mathcal{B}^+$. Now let f be any function of $\mathcal{B}^+$ and choose $g \in U$ (by the lemma) such that $f \leqq g$. There exist $g_n \in L^+$ such that $g_n \uparrow g$. Then $f \cap g_n \in \mathfrak{F}$ (being L-bounded) and $f \cap g_n \uparrow f$, so that $f \in \mathfrak{F}$ by the definition of an L-monotone family. We have proved that $\mathcal{B}^+ \subset \mathfrak{F}$. Since $\mathcal{B}^+$ is itself an L-bounded family including L^+ and since $\mathfrak{F}$ is the smallest such it follows that $\mathcal{B}^+ = \mathfrak{F}$, q.e.d.

12J. We now replace L^1 by $L^1 \cap \mathcal{B}$. That is, from now on a function is not considered to be summable unless it satisfies the earlier definition (**12E**) and also is a Baire function. This restriction to Baire functions is entirely a matter of convenience. It avoids the necessity of certain "measure zero" arguments in proofs such as those of **16C, 31A** and **33A,** and is helpful in situations involving more than one integral.

Theorem. *In order that $f \in L^1$, it is necessary and sufficient that $f \in \mathcal{B}$ and that there exist $g \in L^1$ such that $|f| \leqq g$.*

Proof. The necessity is trivial. In proving the sufficiency we may suppose that $f \geqq 0$. Then the family of functions $h \in \mathcal{B}^+$

such that $h \cap g \in L^1$ is monotone (by **12G**) and includes L^+, and is therefore equal to $\mathfrak{G}^+$. Thus $f = f \cap g \in L^1$, q.e.d.

12K. We now extend I to any function in $\mathfrak{G}^+$ by putting $I(f) = \infty$ if f is not summable. A function $f \in \mathfrak{G}$ will be said to be *integrable* if either its positive part $f^+ = f \cup 0$ or its negative part $f^- = -(f \cap 0)$ is summable. Then $I(f) = I(f^+) - I(f^-)$ is unambiguously defined, although it may have either $+\infty$ or $-\infty$ as its value, and f is summable if and only if it is integrable and $|I(f)| < \infty$. Theorems **12F** and **12G** have the following theorem as an immediate corollary.

Theorem. *If f and g are integrable, then $f + g$ is integrable and $I(f + g) = I(f) + I(g)$, provided that $I(f)$ and $I(g)$ are not oppositely infinite. If f_n is integrable, $I(f_1) > -\infty$ and $f_n \uparrow f$, then f is integrable and $I(f_n) \uparrow I(f)$.*

§ 13. EQUIVALENCE AND MEASURABILITY

13A. If $A \subset S$, then "φ_A" conventionally designates the characteristic function of A: $\varphi_A(p) = 1$ if $p \in A$ and $\varphi_A(p) = 0$ if $p \in A'$. If $\varphi_A \in \mathfrak{G}$, we shall say that the set A is *integrable* and define its *measure* $\mu(A)$ as $I(\varphi_A)$. It follows from **12F** that, if A and B are integrable, then so are $A \cup B$, $A \cap B$, and $A - B$; and from **12G** that, if $\{A_n\}$ is a disjoint sequence of integrable sets, then $\bigcup_1^\infty A_n$ is integrable and $\mu(\bigcup_1^\infty A_n) = \sum_1^\infty \mu(A_n)$. If A is integrable and $\mu(A) < \infty$, then A will be said to be *summable*. The integral I will be said to be *bounded* if S is summable: it is clear that I is then a bounded linear functional with respect to the uniform norm.

13B. We shall now further restrict L by adding a hypothesis used by Stone [47], namely, that

$$f \in L \Rightarrow f \cap 1 \in L.$$

Then also $f \cup (-1) \in L$, and these properties are preserved through the extension, so that $f \in \mathfrak{G} \Rightarrow f \cap 1 \in \mathfrak{G}$. This axiom is added to enable us to prove the following theorem.

Theorem. *If $f \in \mathfrak{G}$ and $a > 0$, then $A = \{p : f(p) > a\}$ is an integrable set. If $f \in L^1$, then A is summable.*

Proof. The function $f_n = [n(f - f \cap a)] \cap 1$ belongs to $\mathfrak{B}$ and it is easy to see that $f_n \uparrow \varphi_A$, so that $\varphi_A \in \mathfrak{B}$ and A is integrable. The second statement follows from the inequality $0 \leqq \varphi_A \leqq f^+/a$.

13C. It is extremely important that the converse of this theorem is true.

Theorem. *If* $f \geqq 0$ *and* $A = \{p : f(p) > a\}$ *is integrable for every positive* a, *then* $f \in \mathfrak{B}$.

Proof. Given $\delta > 1$, let $A_m{}^\delta = \{p : \delta^m < f(p) \leqq \delta^{m+1}\}$, $-\infty < m < \infty$. Let $\varphi_m{}^\delta$ be the characteristic function of $A_m{}^\delta$, and let $f_\delta = \sum_{-\infty}^{\infty} \delta^m \varphi_m{}^\delta$. Then $\varphi_m{}^\delta \in \mathfrak{B}^+$ by hypothesis, and therefore $f_\delta \in \mathfrak{B}^+$. If $\delta \downarrow 1$ through a suitable sequence of values (say $\delta_n = 2^{2^{-n}}$), then $f_\delta \uparrow f$ and it follows that $f \in \mathfrak{B}$, q.e.d.

Corollary 1. *If* $f \in \mathfrak{B}^+$ *and* $a > 0$, *then* $f^a \in \mathfrak{B}^+$.

Proof. $f^a > b$ if and only if $f > b^{1/a}$, and the corollary follows from the theorem and **13B**.

Corollary 2. *If* f *and* $g \in \mathfrak{B}^+$, *then* $fg \in \mathfrak{B}^+$.

Proof. $fg = [(f + g)^2 - (f - g)^2]/4$.

Corollary 3. *If* $f \in \mathfrak{B}^+$, *then* $I(f) = \int f \, d\mu$, *where the integral is taken in the customary sense.*

Proof. The function f_δ in the proof of the theorem was defined with this corollary in mind. We have $f_\delta \leqq f \leqq \delta(f_\delta)$ and $I(f_\delta)$ $= \sum \delta^m I(\varphi_m{}^\delta) = \sum \delta^m \mu(A_m{}^\delta) = \int f_\delta \, d\mu$. Since $I(f_\delta) \leqq I(f) \leqq$ $\delta I(f_\delta)$ and $\int f_\delta \, d\mu \leqq \int f \leqq \delta \int f_\delta \, d\mu$, we see that, if either of the numbers $I(f)$ and $\int f \, d\mu$ is finite, then they both are, and that $|\, I(f) - \int f \, d\mu \,| \leqq (\delta - 1)I(f_\delta) \leqq (\delta - 1)I(f)$. Since δ is any number greater than one, the corollary follows.

13D. A function f is said to be a *null* function if $f \in \mathfrak{B}$ and $I(|\, f \,|) = 0$. A set is a *null* set if its characteristic function is a null function, that is, if it is integrable and its measure is zero. Any integrable subset of a null set is null, and a countable union of null sets is a null set. Any function of $\mathfrak{B}$ dominated by a null function is a null function, and a countable sum of null functions is null. Two functions f and g are said to be *equivalent* if $(f - g)$ is a null function.

Theorem. *A function $f \in \mathfrak{B}$ is null if and only if $\{p : f(p) \neq 0\}$ is a null set.*

Proof. Let $A = \{p : f(p) \neq 0\}$. If A is a null set, then $n\varphi_A$ is a null function and $n\varphi_A \uparrow \infty\varphi_A$, so that the function equal to $+\infty$ on A and zero elsewhere ($\infty\varphi_A$) is a null function. But $0 \leq |f| \leq \infty\varphi_A$ so that f is null. Conversely, if f is null, then so is $(n|f|) \cap 1$, and, since $(n|f|) \cap 1 \uparrow \varphi_A$, it follows that φ_A is a null function and A is a null set.

We are now in a position to remark on an ambiguity remaining in **12F**, namely, that, since the values $\pm\infty$ are allowed for a summable function, the sum $f_1 + f_2$ of two summable functions is not defined on the set where $f_1 = \pm\infty$ and $f_2 = \mp\infty$. However, it is easy to see that, if f is summable, then the set A where $|f| = \infty$ is null, and we can therefore eliminate $\pm\infty$ as possible values if the convergence theorems are restricted to almost everywhere convergence (pointwise convergence except on a null set). The other method of handling this ambiguity is to lump together as a single element all functions equivalent to a given function and then define addition for two such equivalence classes by choosing representatives that do not assume the values $\pm\infty$.

13E. Very often the space S is not itself an integrable set, a fact which is the source of many difficulties in the more technical aspects of measure theory. It may happen, however, and does in the case of the Haar measure on a locally compact group, that S is the union of a (perhaps uncountable) disjoint family $\{S_\alpha\}$ of integrable sets with the property that *every* integrable set is included in an at most countable union of the sets S_α. Now the above-mentioned difficulties do not appear. A set A is defined to be *measurable* if the intersections $A \cap S_\alpha$ are all integrable, and a function is measurable if its restriction to S_α is integrable for every α. It follows that the intersection of a measurable set with *any* integrable set is integrable, and that the restriction of a measurable function to any integrable set is integrable. The notion of a null function can be extended to cover functions whose restrictions to the sets S_α are all null, and the same for null sets. Two functions are *equivalent* if their difference is null in the above sense. The families of measurable functions and of measurable sets have much the same elementary combinatorial properties as the corresponding integrable families and there seems little need for further discussion on this point. The important property of a measurable space of the above kind is the possibility of building up a measurable function by defining it on each of the sets S_α. In a general measure space S there is no way of defining a measurable function by a set of such local definitions.

Since every integrable set is a countable union of summable sets, the sets S_α may be assumed to be summable if so desired.

§ 14.   THE REAL L^p-SPACES

14A. If $p \geqq 1$, the set of functions $f \in \mathfrak{B}$ such that $|f|^p$ is summable is designated L^p. The next few paragraphs will be devoted to showing that L^p is essentially a Banach space under the norm $\| f \|_p = [I(|f|^p)]^{1/p}$. Since $\| f \|_p = 0$ if f is null, it is, strictly speaking, the quotient space $L^p/\mathfrak{N}$ ($\mathfrak{N} =$ the subspace of null functions) that is a Banach space. However, this logical distinction is generally slurred over, and L^p is spoken of as a Banach space of functions.

If $fg \in L^1$, it is convenient to use the scalar product notation $(f, g) = I(fg)$. The Hölder inequality, proved below, shows that L^2 is a real Hilbert space, and it follows similarly that the complex L^2 space discussed later is a (complex) Hilbert space. These spaces are among the most important realizations of Hilbert space.

14B. (Hölder's inequality.)   *If $f \in L^p$, $g \in L^q$ and $(1/p) + (1/q) = 1$, then $fg \in L^1$ and*

$$| (f, g) | \leqq \| f \|_p \| g \|_q.$$

Proof. From the ordinary mean value theorem of the calculus it follows that, if $x \geqq 1$ and $p \geqq 1$, then $x^{1/p} \leqq x/p + 1/q$. If $x = a/b$, this becomes $a^{1/p} b^{1/q} \leqq \dfrac{a}{p} + \dfrac{b}{q}$. If $f \in L^p$, $g \in L^q$, and

$|| f ||_p \neq 0 \neq || g ||_q$, then we can take $a = | f |^p / || f ||_p{}^p$, $b = | g |^q / || g ||_q{}^q$ in this inequality. It follows from **13C** and **12J** that the left member is summable, and integrating we get $I(| fg | / (|| f ||_p \cdot || g ||_q)) \leqq 1$. Hölder's inequality now follows from $| I(h) | \leqq I(| h |)$. If $|| f ||_p = 0$ or $|| g ||_q = 0$, then fg is null and the inequality is trivial.

Corollary. *If* $f, g \in \mathscr{B}^+$, *then* $(f, g) \leqq || f ||_p || g ||_q$.

14C. (Minkowski's inequality.) *If* f *and* $g \in L^p$ $(p \geqq 1)$, *then* $f + g \in L^p$ *and*

$$|| f + g ||_p \leqq || f ||_p + || g ||_p.$$

Proof. The case $p = 1$ follows directly from $| f + g | \leqq | f | + | g |$. If $p > 1$ and $f, g \in L^p$, the inequality $| f + g |^p \leqq [2 \max (| f |, | g |)]^p \leqq 2^p (| f |^p + | g |^p)$ shows that $f + g \in L^p$. Then

$$|| f + g ||_p{}^p \leqq I(| f + g |^{p-1} | f |) + I(| f + g |^{p-1} | g |)$$
$$\leqq || f + g ||_p{}^{p-1} || f ||_p + || f + g ||_p{}^{p-1} || g ||_p,$$

where the second inequality follows from the Hölder inequality. Minkowski's inequality is obtained upon dividing by $|| f + g ||_p{}^{p-1}$. If $|| f + g ||_p = 0$, the inequality is trivial.

Corollary. *If* $f, g \in \mathscr{B}^+$, *then* $|| f + g ||_p \leqq || f ||_p + || g ||_p$.

14D. Since the homogeneity property $|| \lambda f ||_p = | \lambda | \, || f ||_p$ is obvious from the definition of $|| f ||_p$, Minkowski's inequality shows that L^p (actually L^p/null functions) is a normed linear space. It remains to be shown that L^p is complete.

Theorem. L^p *is complete if* $p \geqq 1$.

Proof. We first remark that, if $f_n \in L^p, f_n \geqq 0$ and $\sum_1^\infty || f_n ||_p < \infty$, then $f = \sum_1^\infty f_n \in L^p$ and $|| f ||_p \leqq \sum_1^\infty || f_n ||_p$. For $g_n = \sum_1^n f_i \in L^p$ and $|| g_n ||_p \leqq \sum_1^n || f_i ||_p$ by the Minkowski inequality, and since $g_n \uparrow f$ it follows that $f \in L^p$ and $|| f ||_p = \lim || g_n ||_p \leqq \sum_1^\infty || f_n ||_p$, by **12G**.

Now let $\{f_n\}$ by any Cauchy sequence in L^p. We may suppose for the moment, by passing to a subsequence if necessary, that $|| f_{n+1} - f_n ||_p < 2^{-n}$. If $g_n = f_n - \sum_n^\infty | f_{i+1} - f_i |$ and

$h_n = f_n + \sum_n^\infty |f_{i+1} - f_i|$, it follows from our first remark that g_n and $h_n \in L^p$ and that $|| h_n - g_n ||_p < 2^{-n+2}$. Moreover, the sequences g_n and h_n are increasing and decreasing respectively, and if we take f as, say, $\lim g_n$, then $f \in L^p$ and $|| f - f_n ||_p \leqq || h_n - g_n ||_p < 2^{-n+2}$. Our original sequence is thus a Cauchy sequence of which a subsequence converges to f, and which therefore converges to f itself, q.e.d.

14E. We say that a measurable function f is *essentially bounded above* if it is equivalent to a function which is bounded above, and its essential least upper bound is the smallest number which is an upper bound for some equivalent function. The essential least upper bound of $| f |$ is denoted $|| f ||_\infty$; this is the same symbol used elsewhere for the uniform norm, and $|| f ||_\infty$ is evidently the smallest number which is the uniform norm of a function equivalent to f. If L^∞ is the set of essentially bounded measurable functions, equivalent functions being identified, it is clear that L^∞ is a Banach space (in fact, a Banach algebra) with respect to the norm $|| f ||_\infty$. The following theorem can be taken as motivation for the use of the subscript "∞" in $|| f ||_\infty$.

14F. Theorem. *If $f \in L^p$ for some $p > 0$, then $\lim_{q \to \infty} || f ||_q$ exists and equals $|| f ||_\infty$, where ∞ is allowed as a possible value of the limit.*

Proof. If $|| f ||_\infty = 0$, then $|| f ||_q = 0$ for every $q > 0$ and the theorem is trivial. Otherwise, let a be any positive number less than $|| f ||_\infty$ and let $A_a = \{x : | f(x) | > a\}$. Then $|| f ||_q \geqq a[\mu(A_a)]^{1/q}$. Moreover, $0 < \mu(A_a) < \infty$, the first inequality holding because $a < || f ||_\infty$ and the second because $f \in L^p$. Therefore $\underline{\lim}_{q \to \infty} || f ||_q \geqq a$. Since a is any number less than $|| f ||_\infty$, $\underline{\lim}_{q \to \infty} || f ||_q \geqq || f ||_\infty$.

The dual inequality $\overline{\lim}_{q \to \infty} || f ||_q \leqq || f ||_\infty$ is obvious if $|| f ||_\infty = \infty$. Supposing, then, that $|| f ||_\infty < \infty$, we have $| f |^q \leqq | f |^p (|| f ||_\infty)^{q-p}$ and therefore

$$|| f ||_q \leqq (|| f ||_p)^{p/q} (|| f ||_\infty)^{1-p/q},$$

giving the desired inequality $\overline{\lim}_{q \to \infty} || f ||_q \leqq || f ||_\infty$.

14G. Theorem. *L is a dense subset of L^p, $1 \leqq p < \infty$.*

Proof. We first observe that L is dense in L^1 by the very definition of L^1 (via U). Now let f be any non-negative function in L^p. Let $A = \{x : 1/n < f(x) < n\}$ and let $g = f\varphi_A$. Then $f - g \downarrow 0$ as $n \to \infty$ and we choose n so that $\|f - g\|_p < \epsilon/2$. Since $0 \leqq g \leqq n\varphi_A$, it follows from **12J** that $g \in L^1$. If we choose $h \in L^+$ so that $\|h - g\|_1 < (\epsilon/2n)^p$ and $h \leqq n$ (see **13B**), then $\|h - g\|_p \leqq (n^{p-1}\|h - g\|_1)^{1/p} < \epsilon/2$. Thus

$$\|f - h\|_p \leqq \|f - g\|_p + \|g - h\|_p < \epsilon,$$

q.e.d.

§ 15.   THE CONJUGATE SPACE OF L^p

15A. The variations of a bounded functional. We now consider the space L of **12A** as a real normed linear space under the uniform norm $\|f\|_\infty$ (see **13A**).

Theorem. *Every bounded linear functional on L is expressible as the difference of two bounded integrals (non-negative linear functionals).*

Proof. Let F be the given bounded functional, and, if $f \geqq 0$, let $F^+(f) = \text{lub} \{F(g) : 0 \leqq g \leqq f\}$. Then $F^+(f) \geqq 0$ and $|F^+(f)| \leqq \|F\| \cdot \|f\|$. It is also obvious that $F^+(cf) = cF^+(f)$ if $c > 0$. Consider now a pair of non-negative functions f_1 and f_2 in L. If $0 \leqq g_1 \leqq f_1$ and $0 \leqq g_2 \leqq f_2$, then $0 \leqq g_1 + g_2 \leqq f_1 + f_2$ and $F^+(f_1 + f_2) \geqq \text{lub} \, F(g_1 + g_2) = \text{lub} \, F(g_1) + \text{lub} \, F(g_2) = F^+(f_1) + F^+(f_2)$. Conversely, if $0 \leqq g \leqq f_1 + f_2$, then $0 \leqq f_1 \cap g \leqq f_1$ and $0 \leqq g - f_1 \cap g \leqq f_2$, so that $F^+(f_1 + f_2) = \text{lub} \, F(g) \leqq \text{lub} \, F(f_1 \cap g) + \text{lub} \, F(g - f_1 \cap g) \leqq F^+(f_1) + F^+(f_2)$. Thus F^+ is additive on non-negative functions. But now F^+ can be extended to a linear functional on L by the usual definition: $F^+(f_1 - f_2) = F(f_1) - F(f_2)$, where f_1 and f_2 are non-negative. And F^+ is bounded since $|F^+(f)| \leqq F^+(|f|) \leqq \|F\| \cdot \|f\|$.

Now let $F^-(f) = F^+(f) - F(f)$. Since $F^+(f) \geqq F(f)$ if $f \geqq 0$, we see that F^- is also a bounded non-negative linear functional, and $F = F^+ - F^-$, proving the theorem.

15B. An integral J is said to be *absolutely continuous* with respect to an integral I if every I-null set is J-null.

The Radon-Nikodym Theorem. *If the bounded integral J is absolutely continuous with respect to the bounded integral I, then there exists a unique I-summable function f_0 such that ff_0 is I-summable and $J(f) = I(ff_0)$ for every $f \in L^1(J)$.*

Proof. (From [23].) We consider the bounded integral $K = I + J$ and the real Hilbert space $L^2(K)$. If $f \in L^2(K)$, then $f = f \cdot 1 \in L^1(K)$ and

$$| J(f) | \leqq J(| f |) \leqq K(| f |) \leqq \| f \|_2 \cdot \| 1 \|_2$$

by the Schwarz (or Hölder) inequality. Thus J is a bounded linear functional over $L^2(K)$, and there exists by **10G** a unique $g \in L^2(K)$ such that

$$J(f) = (f, g) = K(fg).$$

Evidently g is non-negative (except on a K-null set). The expansion

$$J(f) = K(fg) = I(fg) + J(fg) = I(fg) + I(fg^2) + J(fg^2)$$

$$= \cdots = I(f \sum_1^n g^i) + J(fg^n)$$

shows first that the set where $g \geqq 1$ is I-null (by taking f as its characteristic function) and so J-null. Thus $fg^n \downarrow 0$ almost everywhere if $f \geqq 0$, and since $f \in L^1(J)$, then $J(fg^n) \downarrow 0$ by **12G**. Second, the same expansion shows that, if $f_0 = \sum_1^\infty g^i$, then $ff_0 \in L^1(I)$ and $J(f) = I(ff_0)$, again by **12G**. Taking $f = 1$, it follows in particular that $f_0 \in L^1(I)$. The I-uniqueness of f_0 follows from the K-uniqueness of g and the relation $g = f_0/(1 + f_0)$. Since the integrals $J(f)$ and $I(ff_0)$ are identical on $L^2(K)$ and, in particular, on L, they are identical on $L^1(J)$.

This proof is also valid for the more general situation in which J is not necessarily absolutely continuous with respect to I. We simply separate off the set N where $g \geqq 1$, which is I-null as above but not now necessarily J-null, and restrict f to the complementary set $S - N$.

15C. Theorem. *If I is a bounded integral and F is a bounded linear functional on $L^p(I)$, $1 \leqq p < \infty$, then there exists a unique* *function* $f_0 \in L^q$ *(where* $q = p/(p-1)$ *if* $p > 1$ *and* $q = \infty$ *if* $p = 1$) *such that* $\| f_0 \|_q = \| F \|$ *and*

$$F(g) = (g, f_0) = I(gf_0)$$

for every $g \in L^p$. *That is,* $(L^p)^* = L^q$.

Proof. The variations of F are integrals on L since $f_n \downarrow 0$ implies $\| f_n \|_p \downarrow 0$ and since F^+ and F^- are bounded by $\| F \|$. Since $\| f \|_p = 0$, and therefore $F^+(f) = F^-(f) = 0$, when f is I-null, the variations of F are absolutely continuous with respect to I. Hence there exists by the Radon-Nikodym theorem a summable function f_0 such that

$$F(g) = I(gf_0)$$

for every function g which is summable with respect to both F^+ and F^-, and in particular for every $g \in L^p$. If g is bounded, $0 \leqq g \leqq | f_0 |$ and $p > 1$, then

$$I(g^q) \leqq I(g^{q-1} \operatorname{sgn} f_0 \cdot f_0) \leqq \| F \| \cdot \| g^{q-1} \|_p = \| F \| \cdot [I(g^q)]^{1/p}$$

so that $\| g \|_q \leqq \| F \|$. Since we can choose such functions g_n so that $g_n \uparrow | f_0 |$, it follows from **12G** that $f_0 \in L^q$ and $\| f_0 \|_q \leqq \| F \|$.

The case $p = 1$ is best treated directly. Suppose that $\| f_0 \|_\infty \geqq \| F \| + \epsilon$ ($\epsilon > 0$) and let g be the characteristic function of $A = \{ p : | f_0(p) | \geqq \| F \| + \epsilon/2 \}$. Then $(\| F \| + \epsilon/2)\mu(A) \leqq I(| gf_0 |) = F(g \operatorname{sgn} f_0) \leqq \| F \| \cdot \| g \|_1 = \| F \| \mu(A)$, a contradiction. Therefore $\| f_0 \|_\infty \leqq \| F \|$.

Conversely, we have $| F(g) | \leqq I(| gf_0 |) \leqq \| g \|_p \| f_0 \|_q$ (by the Hölder inequality if $p > 1$), so that $\| F \| \leqq \| f_0 \|_q$. Together with the above two paragraphs this establishes the equality $\| F \| = \| f_0 \|_q$.

The uniqueness of f_0 can be made to follow from **15B**. It can be as easily checked directly, by observing that, if f_0 is not (equivalent to) zero, then F is not the zero functional, so that nonequivalent functions f_0 define distinct functionals F.

15D. The requirement that I be bounded can be dropped from the preceding theorem if $p > 1$, but if $p = 1$ it must be replaced by some condition such as that of **13E**. The discussion follows.

If the domain of F is restricted to the subspace of L^p consisting of the functions which vanish off a given summable set S_1, then the norm of F is decreased if anything. The preceding theorem furnishes a function f_1 defined on S_1. If S_2 and f_2 are a second such pair, then the uniqueness part of **15C** shows that f_1 and f_2 are equivalent on $S_1 \cap S_2$. Suppose now that $p > 1$. Let $b = \text{lub } \{ \| f \|_q : \text{all such } f \}$ and let S_n be a nested increasing sequence such that $\| f_n \|_q \uparrow b$. Then $\{ f_n \}$ is a Cauchy sequence in L^q and its limit f_0 is confined to $S_0 = \bigcup_1^\infty S_n$, with $\| f_0 \|_q = b$. Because of this maximal property of f_0, there can be no non-null f' whose domain S' is disjoint from S_0. Now if $g \in L^p$ and if the set on which $g \neq 0$ is broken up into a countable union of disjoint summable sets A_m, then the restriction f_m of f_0 to A_m is the function associated by **15C** with A_m, and $F(g) = \sum_1^\infty F(g_m) = \sum_1^\infty I(g_m f_m) = I(g f_0)$. As before, $\| F \| \leq \| f_0 \|_q$, and since $\| f_0 \|_q = b \leq \| F \|$ the equality follows.

Now if $p = 1$ this maximizing process cannot be used to furnish a single function $f_0 \in L^\infty$ and there is in general *no* way of piecing together the functions f_α associated by **15C** with summable sets S_α to form a function f_0 defined on the whole of S. However, if there is a basic *disjoint* family $\{ S_\alpha \}$ with the property discussed in **13E**, then the function f_0 defined to be equal to f_α on S_α is measurable and is easily seen (as in the case $p > 1$ considered above) to satisfy $F(g) = I(g f_0)$ for all $g \in L^1$.

§ 16. INTEGRATION ON LOCALLY COMPACT HAUSDORFF SPACES

16A. We specialize the considerations of § 12 now to the case where S is a locally compact Hausdorff space and L is the algebra of all real-valued continuous functions which vanish off compact sets. We write L^+ for the set of non-negative functions in L, and L_A and L_A^+ for the sets of functions in L and L^+ respectively which vanish outside of A.

Lemma 1. *If $f_n \in L$ and $f_n \downarrow 0$, then $f_n \downarrow 0$ uniformly.*

Proof. Given ϵ, let $C_n = \{ p : f_n(p) \geq \epsilon \}$. Then C_n is compact and $\bigcap_n C_n = \varnothing$, so that $C_n = \varnothing$ for some n. That is, $\| f_n \|_\infty \leq \epsilon$ for some, and so for all later, n, q.e.d.

Lemma 2. *A non-negative linear functional is bounded on L_C whenever C is compact.*

Proof. We can choose $g \in L^+$ such that $g \geqq 1$ on C. Then $f \in L_C$ implies $|f| \leqq \|f\|_\infty g$ and $|I(f)| \leqq I(g) \cdot \|f\|_\infty$, so that $\|I\| \leqq I(g)$ on L_C.

Theorem. *Every non-negative linear functional on L is an integral.*

Proof. If $f_n \downarrow 0$, then $\|f_n\|_\infty \downarrow 0$ by Lemma 1. If $f_1 \in L_C$, then $f_n \in L_C$ for all n, and, if B is a bound for I on C (Lemma 2), then $|I(f_n)| \leqq B \|f_n\|_\infty$ and $I(f_n) \downarrow 0$. Therefore, I is an integral.

16B. *Let S_1 and S_2 be locally compact Hausdorff spaces and let I and J be non-negative linear functionals on $L(S_1)$ and $L(S_2)$ respectively. Then $I_x(J_y f(x, y)) = J_y(I_x f(x, y))$ for every f in $L(S_1 \times S_2)$, and this common value is an integral on $L(S_1 \times S_2)$.*

Proof. Given $f(x, y) \in L(S_1 \times S_2)$, let C_1 and C_2 be compact sets in S_1 and S_2 respectively such that f vanishes off $C_1 \times C_2$, and let B_1 and B_2 be bounds for the integrals I and J on C_1 and C_2 respectively. Given ϵ, we can find a function of the special form $k(x, y) = \sum_1^n g_i(x) h_i(y)$, $g_i \in L_{C_1}$, $h_i \in L_{C_2}$, such that $\|f - k\|_\infty < \epsilon$, for by the Stone-Weierstrass theorem the algebra of such functions k is dense in $L_{C_1 \times C_2}$. It follows, first, that $|J_y f(x, y) - \sum J(h_i) g_i(x)| = |J_y(f - k)| < \epsilon B_2$, so that $J_y f$ is a uniform limit of continuous functions of x and is itself therefore continuous, and, second, that $|I_x J_y f - \sum I(g_i) J(h_i)| < \epsilon B_1 B_2$. Together with the same inequality for $J_y I_x f$, this yields $|I_x J_y f - J_y I_x f| < 2\epsilon B_1 B_2$ for every ϵ, so that $I_x J_y f = J_y I_x f$. It is clear that this functional is linear and non-negative, and therefore an integral.

16C. (Fubini Theorem.) *If K is the functional on $L(S_1 \times S_2)$ defined above and if $f(x, y) \in \mathcal{B}^+(S_1 \times S_2)$, then $f(x, y) \in \mathcal{B}^+(S_1)$ as a function of x for every y, $I_x f(x, y) \in \mathcal{B}^+(S_2)$ and*

$$Kf = J_y(I_x f(x, y)).$$

Proof. We have seen in **16B** that the set $\mathfrak{F}$ of functions of $B^+(S_1 \times S_2)$ for which the theorem is true includes $L^+(S_1 \times S_2)$.

It is also L-monotone. For suppose that $\{f_n\}$ is a sequence of L-bounded functions of $\mathfrak{F}$ and that $f_n \uparrow f$ or $f_n \downarrow f$. Then $K(f)$ $= \lim K(f_n) = \lim J_y(I_x f_n) = J_y(\lim I_x f_n) = J_y(I_x \lim f_n) = J_y I_x f$, by repeated application of 12G or 12K (or their negatives). Thus $\mathfrak{F}$ is L-monotone and includes L^+, and is therefore equal to $\mathfrak{B}^+$ by 12I.

The Fubini theorem is, of course, valid in the absence of topology, but the proof is more technical, and the above case is all that is needed in this book.

16D. The more usual approach to measure and integration in Cartesian product spaces starts off with an elementary measure $\mu(A \times B) = \mu_1(A)\mu_2(B)$ on "rectangles" $A \times B$ $(A \subset S_1, B \subset S_2)$. This is equivalent to starting with an elementary integral

$$\int \sum c_i \, \varphi_{A_i \times B_i} = \sum c_i \, \mu_1(A_i)\mu_2(B_i)$$

over linear combinations of characteristic functions of rectangles. The axioms of 12A for an elementary integral are readily verified. We first remark that for a function $f(x, y)$ of the above sort the definition of $\int f$ is actually in terms of the iterated integral:

$$\int f = \int \left(\int f \, dx \right) dy = \int \left(\int f \, dy \right) dx.$$ If now $f_n(x, y)$ is a sequence of elementary functions and $f_n \downarrow 0$, then $f_n \downarrow 0$ as a function of x for every y and therefore $\int f_n \, dx \downarrow 0$ for every y by the property 12G of the integral in the first measure space S_1. But then $\int \left(\int f_n \, dx \right) dy \downarrow 0$ by the same property of the integral in the second space. Hence $\int f_n \, d\mu \downarrow 0$, proving (4) of 12A. The other axioms are more easily checked and we can therefore assume the whole theory of measure and integration as developed in this chapter.

In case S_1 and S_2 are locally compact spaces and μ_1 and μ_2 arise from elementary integrals on continuous functions, we now have two methods for introducing integration in $S_1 \times S_2$ and it must be shown that they give the same result. If $\mathfrak{B}$ and $\mathfrak{B}'$ are the two families of Baire functions thereby generated ($\mathfrak{B}$ by continuous elementary functions and $\mathcal{B}'$ by the above step functions), it must be shown that $\mathcal{B} = \mathcal{B}'$ and that the two integrals agree on any generating set for this family. Now it is easy to see that $\varphi_{A \times B} \in \mathcal{B}$ if A and B are integrable sets in S_1 and S_2, for $\varphi_{A \times B} = \varphi_A(x)\varphi_B(y)$ lies in the monotone family generated by functions $f(x)g(y)$, where f and g are elementary continuous functions on S_1 and S_2. Thus $\mathcal{B}' \subset \mathcal{B}$. Conversely, if $f \in L(S_1)$ and $g \in L(S_2)$, then $f(x)g(y)$ can be uniformly approximated by a step function of the form $(\sum \varphi_{A_i}(x)(\sum \varphi_{B_j}(y)))$, and, therefore, by the Stone-Weierstrass theorem any continuous function $f(x, y)$ which vanishes off a compact set can be similarly approximated. Thus $\mathcal{B} \subset \mathcal{B}'$, proving that $\mathcal{B} = \mathcal{B}'$. By the same argument the two integrals agree for functions $f(x, y)$ of the form $f(x)g(y)$ and therefore on the whole of $\mathcal{B} = \mathcal{B}'$.

§ 17.   THE COMPLEX L^p-SPACES

17A. We make the obvious definition that a complex-valued function $f = u + iv$ is measurable or summable if and only if its real and imaginary parts u and v are measurable or summable. Evidently a measurable function f is summable if and only if $|f| = (u^2 + v^2)^{1/2}$ is summable, and we can prove the critical inequality:

$$| I(f) | \leqq I(| f |).$$

Proof. We observe that $h = u^2/(u^2 + v^2)^{1/2} \in L^1$. The Schwarz inequality can therefore be applied to $| u | = h^{1/2}(u^2 + v^2)^{1/4}$ giving $| I(u) |^2 \leqq I(h)I((u^2 + v^2)^{1/2})$. Writing down the corresponding inequality for v and adding, we obtain

$$| I(u) |^2 + | I(v) |^2 \leqq [I((u^2 + v^2)^{1/2})]^2,$$

as was desired.

17B. The definition of L^p is extended to include all complex-valued measurable functions f such that $| f |^p$ is summable with $\| f \|_p = [I(| f |^p)]^{1/p}$ as before, and we see at once that $f \in L^p$ if and only if $u, v \in L^p$. The Hölder and Minkowski inequalities remain unchanged by virtue of the inequality in **17A**. The extended L^p is complete, for a sequence $f_n = u_n + iv_n$ is Cauchy if and only if its real and imaginary parts u_n and v_n are real Cauchy sequences.

17C. A bounded linear functional F on L^p, considered on the subset of real-valued functions, has real and imaginary parts each of which is a real-valued bounded linear functional. Taken together these determine by **15C** a complex-valued measurable function $f_0 \in L^q$ such that $F(g) = I(g\bar{f}_0)$ for every real function $g \in L^p$, and hence, because F is complex-homogeneous, for every $g \in L^p$. The proof that $\| F \| = \| f_0 \|_q$ holds as well in the complex case as in the real case if we replace f_0 by $\bar{f}_0$ throughout **15C** and remember that $\operatorname{sgn} f_0 = e^{i \arg f_0}$.

Since this time we already know that $f_0 \in L^q$ a slight simplification can be made, as follows. That $\| F \| \leq \| f_0 \|_q$ follows, as before, from the Hölder inequality. If $p > 1$, we can take $g = | f_0 |^{q-1} e^{i \arg f_0}$ and get $F(g) = I(| f_0 |^q) = \| f_0 \|_q{}^q$, $\| g \|_p = (\| f_0 \|_q)^{q/p}$ and $| F(g) | / \| g \|_p = \| f_0 \|_q$. Thus $\| F \| \geq \| f_0 \|_q$, and so $\| F \| = \| f_0 \|_q$.

Chapter IV

BANACH ALGEBRAS

This chapter is devoted to an exposition of part of the theory of Banach algebras, with emphasis on the commutative theory stemming from the original work of Gelfand [12]. This theory, together with its offshoots, is having a marked unifying influence on large sections of mathematics, and in particular we shall find that it provides a basis for much of the general theory of harmonic analysis. We begin, in section 19, with a motivating discussion of the special Banach algebra $\mathcal{C}(S)$ of all continuous complex-valued functions on a compact Hausdorff space, and then take up the notions of maximal ideal, spectrum and adverse (or quasi-inverse) thus introduced. The principal elementary theorems of the general theory are gathered together in § 24.

§ 18. DEFINITION AND EXAMPLES

Definition. *A Banach algebra is an algebra over the complex numbers, together with a norm under which it is a Banach space and which is related to multiplication by the inequality:*

$$\| xy \| \leqq \| x \| \, \| y \|.$$

If a Banach algebra A has an identity e, then $\| e \| = \| ee \| \leqq \| e \|^2$, so that $\| e \| \geqq 1$. Moreover, it is always possible to renorm A with an equivalent smaller norm $\| \| x \| \|$ so that $\| \| e \| \| = 1$; we merely give an element y the norm of the bounded linear transformation which left-multiplication by y defines on A ($\| \| y \| \| = \mathrm{lub} \, \| yx \| / \| x \|$) and observe that then $\| y \| / \| e \| \leqq$

$||| \, y \, ||| \leq \, || \, y \, ||$. We shall always suppose, therefore, that $|| \, e \, ||$ $= 1$.

If A does not have an identity, it will be shown in **20C** that the set of ordered pairs $\{ \langle x, \lambda \rangle \colon \lambda$ is a complex number and $x \in A \}$ is an extension of A to an algebra with an identity, and a routine calculation will show the reader that the norm $|| \, \langle x, \lambda \rangle \, || = || \, x \, || + | \, \lambda \, |$ makes the extended algebra into a Banach algebra.

We have already pointed out several Banach algebras in our preliminary chapters. These were:

1a. The bounded continuous complex-valued functions on a topological space S, with the uniform norm $|| \, f \, ||_\infty = \mathrm{lub}_{x \in S} | \, f(x) \, |$ (see **4B**).

1b. The essentially-bounded measurable complex-valued functions on a measure space S, with the essential uniform norm (see **14E**).

2a. The bounded linear transformations (operators) on a Banach space (see **7C**).

2b. The bounded operators on a Hilbert space (§ 11).

Of these algebras, 1a and 2b will be of continued interest to us. Still more important will be the group algebras of locally compact groups, of which the following are examples:

3a. The sequences of complex numbers $a = \{ a_n \}$ with $|| \, a \, || = \sum_{-\infty}^{\infty} | \, a_n \, | < \infty$ and with multiplication $a * b$ defined by

$$(a * b)_n = \sum_{m = -\infty}^{m = +\infty} a_{n-m} b_m.$$

Here we have the group G of integers as a measure space, the measure of each point being 1. The norm is simply the L^1 norm, $|| \, a \, || = \int_G | \, a \, |$, and the algebra is $L^1(G)$. Its multiplication as defined above is called *convolution*. We shall omit checking the algebraic properties of convolution, but it should be noticed that the norm inequality $|| \, a * b \, || \leq || \, a \, || \, || \, b \, ||$ arises from the reversal of a double summation, which is a special case of the Fubini theorem:

$$|| \, a * b \, || = \sum_n | \, \sum_m a_{n-m} b_m \, | \leq \sum_m (\sum_n | \, a_{n-m} b_m \, |)$$
$$= \sum_m | \, b_m \, | \, || \, a \, || = || \, a \, || \, || \, b \, ||.$$

It is clear that the sequence e defined by $e_0 = 1$ and $e_n = 0$ if $n \neq 0$ is an identity for this algebra.

3b. $L^1(-\infty, \infty)$ with convolution as multiplication:

$$[f * g](x) = \int_{-\infty}^{\infty} f(x - y)g(y)\, dy$$

Here the group is the additive group of the real numbers, with ordinary Lebesgue measure. The inequality $\| f * g \|_1 \leqq \| f \|_1 \| g \|_1$ is again due to the Fubini theorem:

$$\| f * g \|_1$$

$$= \int_{-\infty}^{\infty} \left| \int_{-\infty}^{\infty} f(x - y)g(y)\, dy \right| dx \leqq \int_{-\infty}^{\infty} \int_{-\infty}^{\infty} | f(x - y)g(y) | \, dx\, dy$$

$$= \| f \|_1 \| g \|_1.$$

In each of 3a and 3b the measure used is the so-called Haar measure of the group.

The following are examples of important types of Banach algebras with which we shall be less concerned.

4. The complex-valued functions continuous on the closed unit circle $\| z \| \leqq 1$ and analytic on its interior $\| z \| < 1$, under the uniform norm.

5. The complex-valued functions on $[0, 1]$ having continuous first derivatives, with $\| f \| = \| f \|_\infty + \| f' \|_\infty$.

§ 19. FUNCTION ALGEBRAS

The imbedding of a Banach space X in its second conjugate space X^{**}, that is, the representation of X as a space of linear functionals over the Banach space X^*, is an important device in the general theory of Banach spaces. In Gelfand's theory of a commutative Banach algebra A the corresponding representation is *all*-important. We consider now the space Δ of all continuous *homomorphisms* of A onto the complex numbers, for every $x \in A$ the function $\hat{x}$ on Δ defined by $\hat{x}(h) = h(x)$ for all $h \in \Delta$, and the algebraic *homomorphism* of A onto the algebra $\hat{A}$ of all such functions $\hat{x}$. It is clear that Δ is a subset of the conjugate space A^* of A considered as a Banach space, and the function $\hat{x}$ is obtained by taking the functional x^{**}, with domain A^*, and restricting its domain to Δ. This change of domain greatly changes the nature of the theory. For example, Δ is a weakly closed subset of the closed unit sphere in A^* and is therefore weakly compact. Moreover, the linear character of x^{**} becomes obliterated, since Δ has no algebraic properties, and $\hat{x}$ is merely a continuous function on a compact space. The more nearly analogous representation theorem for Banach spaces would be the fact that, if x^0 is the restriction of x^{**} to the strongly closed unit sphere in A^*, then the mapping $x \to x^0$ is a linear isometry of X with a Banach space (under the uniform norm) of continuous complex-valued functions on a compact Hausdorff space. But, whereas this representation theorem for a Banach space is devoid of implication and is of the nature of a curio, the corresponding representation $x \to \hat{x}$ of a commutative Banach algebra by an algebra of functions is of the utmost significance. Thus the representation algebra $\hat{A}$ is the vehicle for the development of much of the theory of Banach algebras, and the transformation $x \to \hat{x}$ is a generalization of the Fourier transform.

If our sole concern were the commutative theory we would take the simple concept of a function algebra as central and let the theory grow around it. We shall not adopt this procedure because we want to include a certain amount of general noncommutative theory. However, we shall start out in this section with some simple observations about function algebras, partly as motivation for later theory and partly as first steps in the theory itself.

19A. A homomorphism h of one complex algebra onto another is, of course, a ring homomorphism which also preserves scalars: $h(\lambda x) = \lambda h(x)$. The statements in the following lemma are obvious.

Lemma. *Let $x \to \hat{x}$ be a homomorphism of an algebra A onto an algebra $\hat{A}$ of complex-valued functions on a set S. If p is a point of S, then the mapping $x \to \hat{x}(p)$ is a homomorphism h_p of A into the complex numbers whose kernel is either a maximal ideal M_p of deficiency 1 in A or is the whole of A. The mapping $p \to h_p$ imbeds S (possibly in a many-to-one way) in the set H of all homo-*

morphisms of A into the complex numbers. Conversely, if S is any subset of H and if the function $\hat{x}$ is defined on S by $\hat{x}(h) = h(x)$ for every $h \in S$, then the mapping $x \to \hat{x}$ is a homomorphism of A onto the algebra $\hat{A}$ of all such functions.

A pseudo-norm is a non-negative functional which satisfies all the requirements of a norm except that there may exist non-zero x such that $\| x \| = 0$. A pseudo-normed algebra has the obvious definition. The inequalities $\| x + y \| \leq \| x \| + \| y \|$ and $\| xy \| \leq \| x \| \| y \|$ show that the set I of all x such that $\| x \| = 0$ is an ideal in I. Moreover $\| x \|$ is constant on each coset of the quotient algebra A/I and defines a proper norm there. Any linear mapping with domain A which is bounded with respect to the pseudo-norm is zero on I and therefore by **7D** is transferable to the domain A/I without changing its norm. In particular the conjugate space A^* is thus identified with $(A/I)^*$.

Lemma. *If the algebra $\hat{A}$ in the above lemma consists of bounded functions, then $\hat{A}$ defines a pseudo norm in A, $\| x \| = \| \hat{x} \|_\infty$, and the mapping $p \to h_p$ identifies S with a subset of the unit sphere in the conjugate space A^* of A.*

Proof. Obvious.

19B. The following theorem is very important for our later work. It is our replacement for the weak compactness of the unit sphere in A^*.

Theorem. *If A is a normed algebra and if Δ is the set of all continuous homomorphisms of A onto the complex numbers, then Δ is locally compact in the weak topology defined by the functions of $\hat{A}$. If A has an identity, then Δ is compact. If Δ is not compact, then the functions of $\hat{A}$ all vanish at infinity.*

Proof. If $h \in \Delta$, then its nullspace M_h is a closed ideal and the quotient space is a normed linear space isomorphic to the complex number field. If E is the identity coset, then $h(x) = \lambda$ if and only if $x \in \lambda E$. Now E cannot contain elements of norm less than 1 since otherwise, being closed under multiplication, it would contain elements of arbitrarily small norm. In general, if $x \in \lambda E$ then $x/\lambda \in E$, $\| x/\lambda \| \geq 1$ and $\| x \| \geq | \lambda | = | h(x) |$.

Thus $\| h \| \leqq 1$ for every $h \in \Delta$, and Δ is a subset of the unit sphere S_1 of A^*. We know S_1 to be weakly compact (**9B**) and therefore the closure $\bar{\Delta}$ of Δ in S_1 is compact.

Now the same argument as in **9B** shows that, if $F \in \bar{\Delta}$, then F is a homomorphism. In fact, if x, y and ϵ are given, there exists by the hypothesis that $F \in \bar{\Delta}$ and the definition of the weak topology an element $h \in \Delta$ such that $| F(x) - h(x) | < \epsilon$, $| F(y) - h(y) | < \epsilon$ and $| F(xy) - h(xy) | < \epsilon$. Since $h(xy) = h(x)h(y)$, it follows that $| F(xy) - F(x)F(y) | < 3\epsilon$ for every ϵ, and therefore that $F(xy) = F(x)F(y)$. Thus either F is a non-zero homomorphism and $F \in \Delta$, or $F = 0$.

We thus have two possibilities: either $\bar{\Delta} = \Delta$ and Δ is compact, which occurs if the zero homomorphism h_0 is not in the closure of Δ, or else $\bar{\Delta} = \Delta \cup \{h_0\}$ and Δ is a compact space minus one point, and hence locally compact. In the second case every function $\hat{x} \in \hat{A}$ vanishes at infinity, for $\hat{x}$ is continuous on the one-point compactification $\bar{\Delta} = \Delta \cup \{h_0\}$ and $\hat{x}(\infty) = h_0(x) = 0$.

If A has an identity e, then $\hat{e}(h) = h(e) = 1$ for all $h \in \Delta$ and it follows that $\hat{e}(F) = F(e) = 1$ for any $F \in \bar{\Delta}$. This rules out the possibility that $h_0 \in \bar{\Delta}$, so that the first case occurs and Δ is compact. This completes the proof of the theorem.

A by-product of the above proof is the fact that $\| \hat{x} \|_\infty \leqq \| x \|$ for every $x \in A$. We thus have a standard norm-decreasing representation of a normed algebra A by an algebra $\hat{A}$ of continuous complex-valued functions on the locally compact Hausdorff space Δ of all continuous homomorphisms of A onto the complex numbers. We shall call this the Gelfand representation of A.

In the simple case of a normed linear space which we sketched earlier the mapping $x \to x^0$ was an isometry roughly because there exist lots of continuous linear functionals (**8C**). Now it is obviously harder for a functional to be a homomorphism than to be merely linear; in fact, there may not exist any non-zero homomorphisms at all. In any case the set Δ is sparse compared to the unit sphere of A^* and we must expect that the Gelfand representation is in general norm *decreasing* and that there may be x such that $\hat{x} \equiv 0$. If the mapping is one-to-one (although probably norm decreasing), A will be called a *function algebra*, and will generally be replaced by the isomorphic algebra of functions $\hat{A}$.

19C. The correspondence $h \to M_h$ between a homomorphism and its kernel allows us to replace Δ by the set $\mathfrak{M}$ of all closed maximal ideals of A which have deficiency 1 as subspaces. In general there will exist many other maximal ideals, some not closed and some not of deficiency 1, which therefore do not correspond to points of Δ. It is very important for the successful application of the methods of algebra that this cannot happen if A is a commutative Banach algebra with an identity; now every maximal ideal is closed and is the kernel of a homomorphism of A onto the complex numbers. This will be shown later for the general case; we prove it here for the special algebra which in the light of the above lemma must be considered as the simplest kind of function algebra, namely, the algebra $\mathcal{C}(S)$ of all continuous complex-valued functions on a compact Hausdorff space S.

Theorem. *If S is a compact Hausdorff space and if I is a proper ideal of $\mathcal{C}(S)$, then there exists a point $p \in S$ such that $I \subset M_p$. If I is maximal, then $I = M_p$. The correspondence thus established between maximal ideals of $\mathcal{C}(S)$ and points of S is one-to-one.*

Proof. Let I be a proper ideal of $\mathcal{C}(S)$ and suppose that for every $p \in S$ there exists $f_p \in I$ such that $f_p(p) \neq 0$. Then $|f_p|^2 = f_p \bar{f}_p \in I$, $|f_p|^2 \geqq 0$ and $|f_p|^2 > 0$ on an open set containing p. By the Heine-Borel theorem there exists a function $f \in I$, a finite sum of such functions $|f_p|^2$, which is positive on S. Then $f^{-1} \in \mathcal{C}(S)$, $1 = f \cdot f^{-1} \in I$ and $\mathcal{C}(S) = I$, a contradiction. Therefore, $I \subset M_p$ for some p, and, if I is maximal, then $I = M_p$. Finally, if $p \neq q$, then there exists $f \in \mathcal{C}(S)$ such that $f(p) = 0$ and $f(q) \neq 0$ (see **3C**) so that $M_p \neq M_q$ and the mapping $p \to M_p$ is one-to-one.

19D. We proved above not only that all the maximal ideals of $\mathcal{C}(S)$ are closed and of deficiency 1 but also that they are all given by points of S. In the general theory of function algebras these two properties can be separated. Thus it follows from simple lemmas proved later (**21D-F**) that, if A is an algebra of bounded functions, then every maximal ideal of A is closed and of deficiency 1 provided A is *inverse-closed*, that is, provided A has the property that whenever $f \in A$ and $\text{glb} |f| > 0$, then

$1/f \in A$. We can then separately raise the question as to when an inverse-closed algebra A of bounded functions on a set S is such that $S = \Delta$. A necessary condition (supposing that A has an identity) is clearly that S be compact under the weak topology defined by A, so that in view of **5G** we may as well suppose that we are given a compact Hausdorff space S and a separating, inverse-closed algebra of continuous functions on S. This is not sufficient, however; we still may have $S \neq \Delta$. If we add one further property to A, the situation suddenly becomes again very simple. We define a function algebra A to be *self-adjoint* if $f \in A \Rightarrow \bar{f} \in A$.

Lemma. *If A is a separating, self-adjoint, inverse-closed algebra of continuous complex-valued functions on a compact space S, then every maximal ideal of A is of the form $M = M_p$ for some $p \in S$. In particular $\Delta = S$.*

Proof. Identical with that of the above theorem in **19C**.

Corollary. *If A is a separating, self-adjoint, inverse-closed algebra of bounded complex-valued functions on a set S, then S is dense in Δ.*

Proof. If $\bar{S}$ is the closure of S in Δ, then $\bar{S}$ is compact and the above lemma applies to the extensions to $\bar{S}$ of the functions of A, proving that $\bar{S} = \Delta$ as required.

A topology $\mathfrak{I}$ on a space S which is completely determined by its continuous complex-valued functions is said to be *completely regular*. By this condition we mean that $\mathfrak{I}$ is identical with the weak topology $\mathfrak{I}_w$ defined on S by the algebra $\mathfrak{C}(S)$ of all bounded complex-valued $\mathfrak{I}$-continuous functions. Since the inclusion $\mathfrak{I}_w \subset \mathfrak{I}$ always holds, complete regularity is equivalent to the inclusion $\mathfrak{I} \subset \mathfrak{I}_w$, and it is easy to see that a necessary and sufficient condition for this to hold is that, given any closed set C and any point $p \notin C$, there exists a continuous real-valued function f such that $f \equiv 0$ on C and $f(p) = 1$.

If S is completely regular, then the above corollary applied to $\mathfrak{C}(S)$ gives an imbedding of S as a dense set of the compact Hausdorff space Δ such that the given topology of S is its relative topology as a subset of Δ. This is a standard compactification of a completely regular space.

19E. The fact that the property proved in **19C** for the algebra $\mathcal{C}(S)$ of all continuous complex-valued functions on a compact Hausdorff space S is now seen (in **19D**) to hold for a much more general class of function algebras suggests the presence of further undisclosed properties of $\mathcal{C}(S)$. We consider two here, starting with a relatively weak property which $\mathcal{C}(S)$ also shares with many other function algebras. In order to present these properties in a form most suitable for later comparison with other algebras we make the following general definitions.

DEFINITION. In any ring with an identity the *kernel* of a set of maximal ideals is the ideal which is their intersection. The *hull* of an ideal I is the set of all maximal ideals which include I.

The reader is reminded of the other conventional use of the word "kernel," as in **19A**.

Theorem. *If $B \subset S$, then $\bar{B} = hull\ (kernel\ (B))$.*

Proof. Let I_B be the kernel of B.

$$I_B = \{f : f \in \mathcal{C}(S) \quad \text{and} \quad f = 0 \text{ on } B\}.$$

If $C = \bar{B}$, then $I_B = I_C$, for a continuous function vanishes on B if and only if it vanishes on $\bar{B} = C$. If p is not in C, there exists by **3C** a continuous function f which vanishes on C ($f \in I_C$) but not at p ($f \notin I_p$). Thus $I_C \subset I_p$ if and only if $p \in C$, and hull $(I_C) = C$. Altogether we have $\bar{B} = C = \text{hull } (I_C) = \text{hull } (I_B)$, q.e.d.

19F. We saw above that a subset of a compact Hausdorff space S is closed if and only if it is equal to the hull of its kernel, when it is considered as a set of maximal ideals in the algebra $\mathcal{C}(S)$. Now this hull-kernel definition of closure can be used to introduce a topology in the space of maximal ideals of any algebra with an identity. This will be discussed further in **20E**; our present concern is the comparison of this hull-kernel topology $\mathfrak{I}_{hk}$, defined on a set S by an algebra A of complex-valued functions on S, with the weak topology $\mathfrak{I}_w$ defined by A, or with any other (stronger) topology $\Im$ in which the functions of A are all continuous. We prove the following theorem.

Theorem. *If A is an algebra of continuous complex-valued functions on a space S with topology $\Im$, then $\Im = \Im_{hk}$ if and only if for every closed set $C \subset S$ and every point p not in C there exists $f \in A$ such that $f \equiv 0$ on C and $f(p) \neq 0$.*

Proof. We first remark that since there may be maximal ideals of A other than those given by points of S it is the intersection with S of the actual hull that is referred to here. Let C be $\Im_{hk}$-closed; that is, $C = \text{hull}\,(I)$ where $I = \text{kernel}\,(C)$. Now the set of maximal ideals containing a given element $f \in A$ is simply the nullspace of the function f and is therefore closed. The hull of I, being the intersection of these nullspaces for all $f \in I$, is therefore closed. Thus every hull-kernel closed set C is also closed and we have the general inclusion $\Im_{hk} \subset \Im$.

The theorem therefore reduces to finding the condition that a closed set C be the hull of its kernel. But the kernel of C is the ideal of elements $f \in A$ such that $f = 0$ on C, and the hull of this ideal will be exactly C if and only if for every $p \notin C$ there exists f in the ideal such that $f(p) \neq 0$. This is the condition of the theorem.

If $A = \mathcal{C}(S)$ in the above theorem, then the condition of the theorem is exactly the one characterizing completely regular spaces. Thus for a completely regular space $\Im = \Im_w = \Im_{hk}$. A function algebra A which satisfies the above condition is called a *regular* function algebra.

19G. The following result is much less susceptible of generalization than that in **19E**. The extent to which it holds or fails to hold in such regular algebras as the algebra of L^1-Fourier transforms over $(-\infty, \infty)$ is related to such theorems as the Wiener Tauberian theorem.

Theorem. *If S is a compact Hausdorff space and I is a uniformly closed ideal in $\mathcal{C}(S)$, then $I = \text{kernel}\,(\text{hull}\,(I))$.*

Proof. Let $C = \text{hull}\,(I)$, $C = \{p : f(p) = 0 \text{ for all } f \in I\}$. Since each $f \in I$ is continuous the null set (hull) of f is closed, and C, the intersection of these null sets, is therefore closed. Let

S_1 be the space $S - C$. Then S_1 is locally compact and the functions of I_C, confined to S_1, form the algebra of all continuous functions which vanish at infinity (see **3D**). The functions of I form a uniformly closed subalgebra of I_C. Let p_1 and p_2 be distinct points of S_1 and let f be a function of $\mathcal{C}(S)$ such that $f = 0$ on C, $f(p_2) = 0$ and $f(p_1) = 1$. Such a function exists by **3C**. Let g be a function of I such that $g(p_1) \neq 0$. Then $gf \in I$, $gf(p_1) \neq 0$ and $gf(p_2) = 0$. Therefore by the Stone-Weierstrass theorem, **4E**, we can conclude that $I = I_C$, q.e.d.

§ 20. MAXIMAL IDEALS

We begin the general theory with some results on maximal ideals in arbitrary rings and algebras. The following simple theorem is perhaps the basic device in the abstract development of harmonic analysis which we are pursuing. Its proof depends explicitly on Zorn's lemma.

20A. Theorem. *In a ring with an identity every proper (right) ideal can be extended to a maximal proper (right) ideal.*

Proof. We consider the family $\mathfrak{F}$ of all proper (right) ideals in the ring R which includes the given ideal I. This family is partially ordered by inclusion. The union of the ideals in any linearly ordered subfamily is an ideal, and is proper since it excludes the identity. Therefore, every linearly ordered subfamily has an upper bound in $\mathfrak{F}$, and $\mathfrak{F}$ contains a maximal element by Zorn's lemma, q.e.d.

20B. If R does not have an identity, the above proof fails because the union of a linearly ordered subfamily of $\mathfrak{F}$ cannot be shown to be a proper ideal. However, the proof can be rescued if R has a (left) identity modulo I, that is, an element u such that $ux - x \in I$ for every $x \in R$, for then u has the necessary property of being excluded from every proper (right) ideal J including I ($u \in J$ and $ux - x \in I \Rightarrow x \in J$ for every $x \in R$, contradicting the assumption that J is proper), and the same proof goes through using u instead of e. A (right) ideal I modulo which R has a (left) identity is said to be *regular*. We have proved:

Theorem. *Every proper regular (right) ideal can be extended to a regular maximal (right) ideal.*

20C. We shall find throughout that the presence of an identity in an algebra makes the theory simpler and more intuitive than is possible in its absence. It is therefore important to observe, as we do in the theorem below, that we can always enlarge an algebra deficient in this respect to one having an identity, and we shall use this device wherever it seems best to do so. However, in many important contexts this extension seems unnatural and undesirable, and we shall therefore carry along a dual development of the theory so as to avoid it wherever feasible.

Theorem. *If A is an algebra without an identity, then A can be imbedded as a maximal ideal of deficiency one in an algebra A_e having an identity in such a way that the mapping $I_e \to I = A \cap I_e$ is a one-to-one correspondence between the family of all (right) ideals I_e in A_e which are not included in A and the family of all regular (right) ideals I of A.*

Proof. The elements of A_e are the ordered pairs $\langle x, \lambda \rangle$, where $x \in A$ and λ is a complex number. Considering $\langle x, \lambda \rangle$ as $x + \lambda e$, the definition of multiplication is obviously $\langle x, \lambda \rangle \langle y, \mu \rangle = \langle xy + \lambda y + \mu x, \lambda \mu \rangle$; we omit the routine check that the enlarged system A_e is an algebra. It is clear that $\langle 0, 1 \rangle$ is an identity for A_e and that the correspondence $x \to \langle x, 0 \rangle$ imbeds A in A_e as a maximal ideal with deficiency 1.

Now let I_e be any (right) ideal of A_e not included in A and let $I = I_e \cap A$. I_e must contain an element v of the form $\langle x, -1 \rangle$. Then the element $u = v + e = \langle x, 0 \rangle \in A$ is a left identity for I_e in A_e ($uy - y = (u - e)y = vy \in I_e$ for all $y \in A_e$) and hence automatically for I in A. Thus I is regular in A. Moreover, since $uy - y \in I_e$ and $uy \in A$ for all y, we see that $y \in I_e$ if and only if $uy \in I$.

Conversely, if I is a regular (right) ideal in A and $u = \langle x, 0 \rangle$ is a left identity for I in A, we define I_e as $\{y : uy \in I\}$. Direct consideration of the definition of multiplication in A_e shows that I is a (right) ideal in A_e; hence I_e is a (right) ideal in A_e. It is not included in A since $u\langle x, -1 \rangle = u(u - e) = u^2 - u \in I$ and therefore $u - e = \langle x, -1 \rangle \in I_e$. Moreover, the fact that $uy - y \in I$ for every $y \in A$ shows that $y \in I$ if and only if $uy \in I$ and $y \in A$, i.e., $I = I_e \cap A$.

We have thus established a one-to-one inclusion preserving correspondence between the family of all regular (right) ideals in A and the family of all (right) ideals of A_e not included in A. In particular the regular maximal (right) ideals of A are the intersections with A of the maximal (right) ideals of A_e different from A.

Corollary. *If S is a locally compact but not compact Hausdorff space and $\mathfrak{C}(S)$ is the algebra of continuous complex-valued functions vanishing at infinity, then the regular maximal ideals of $\mathfrak{C}(S)$ are given by the points of S in the manner of* **19C.**

Proof. The extended algebra is isomorphic to the algebra $C(S_\infty)$ of all continuous complex-valued functions on the one point compactification of S, under the correspondence $\langle f, \lambda \rangle \rightarrow f + \lambda$. In the latter algebra $\mathfrak{C}(S)$ is the maximal ideal corresponding to the point at ∞, and its other maximal ideals, corresponding to the points of S, give the regular maximal ideals of $\mathfrak{C}(S)$ by the above theorem.

20D. *If M is a regular ideal in a commutative ring R, then M is maximal if and only if R/M is a field.*

Proof. If R/M has a proper ideal J, then the union of the cosets in J is a proper ideal of R properly including M. Thus M is maximal in R if and only if R/M has no proper ideals. Therefore, given $X \in R/M$ and not zero, the ideal $\{XY : Y \in R/M\}$ is the whole of R/M. In particular, $XY = E$ for some Y, where E is the identity of R/M. Thus every element X has an inverse and R/M is a field. The converse follows from the fact that a field has no non-trivial ideals.

If R is an algebra over the complex numbers, then the above field is a field over the complex numbers. If R is a Banach algebra, we shall see in § 22 that this field is the complex number field itself, and we shall therefore be able to proceed with our representation program.

20E. We conclude this section with some simple but important properties of the hull-kernel topology, taken largely from Segal [44]. If $\mathfrak{M}$ is the set of regular maximal (two-sided) ideals of a ring R and $B \subset \mathfrak{M}$, we have defined $\bar{B}$ as $h(k(B)) = $ hull (kernel

(B)). In our earlier discussion it was unnecessary to know that the operation $B \to \bar{B}$ was a proper closure operation. Actually, the properties $A \subset B \Rightarrow \bar{A} \subset \bar{B}$ and $B \subset \bar{B} = \bar{\bar{B}}$ follow at once from the obvious monotone properties $A \subset B \Rightarrow k(B) \subset k(A)$, and $I \subset J \Rightarrow h(J) \subset h(I)$. However, the law $\overline{A \cup B} = \bar{A} \cup \bar{B}$ is more sophisticated, and we present a formal proof.

Lemma. *If A and B are closed subsets of $\mathfrak{M}$, then $A \cup B$ is closed.*

Proof. Suppose that $A \cup B$ is not closed. Then there exists $M \in \mathfrak{M}$ such that $k(A \cup B) \subset M$ and $M \not\subset A \cup B$. Since A is closed, $k(A) \not\subset M$ and so $k(A) + M = R$. If e is an identity modulo M, it follows that there exist $a \in k(A)$ and $m_1 \in M$ such that $e = a + m_1$. Similarly, there exist $b \in k(B)$ and $m_2 \in M$ such that $e = b + m_2$. Multiplying we get $e^2 - ab \in M$, and since $e^2 - e \in M$ we see that ab is an identity modulo M. But $ab \in M$ (since $ab \in k(A) \cap k(B) = k(A \cup B) \subset M$), and this is a contradiction.

Remarks: The reader should notice that this proof is not valid in the space of regular maximal right ideals, since we then can conclude neither that $e^2 - ab \in M$ nor that $ab \in M$. The notion of hull-kernel closure is still available, but we cannot conclude that it defines a topology.

We also call explicitly to the reader's attention the fact that every hull is closed. This, like the elementary topological properties listed before the above lemma, is due to the inclusion reversing nature of the hull and kernel mappings. Thus if $C =$ hull I, then $I \subset k(C)$ and $h(k(C)) \subset h(I) = C$, or $\bar{C} \subset C$, as required. Similarly any kernel is the kernel of its hull.

20F. We saw above in **19G** that $I = k(h(I))$ if I is a closed ideal in the algebra $\mathfrak{C}(S)$ of all continuous complex-valued functions on a compact Hausdorff space S. This result does not hold for ideals in general rings, even when they are closed under suitable topologies. The best general result of this nature is the following theorem, various versions of which were discovered independently by Godement, Segal, and Silov. It is the algebraic basis of that part of harmonic analysis exemplified by the Wiener

Tauberian theorem. We define a ring R to be semi-simple if the intersection of its regular maximal ideals is zero.

Lemma. *If I and J are ideals with disjoint hulls and J is regular, then I contains an identity modulo J.*

Proof. An identity e modulo J is also an identity modulo $I + J$, and $I + J$ is therefore regular. But by hypothesis $I + J$ is included in no regular maximal ideal; therefore $I + J = R$, and in particular there exist $i \in I$ and $j \in J$ such that $i + j = e$. The element $i = e - j$ is clearly an identity modulo J, q.e.d.

Theorem. *Let R be a semi-simple ring, I an ideal in R and U an open set of regular maximal ideals such that $h(I) \subset U$ and $k(U')$ is regular. Then $k(U) \subset I$.*

Proof. The hypothesis that U' is closed is equivalent to $U' = h(J)$, where $J = k(U')$. Thus I and J have disjoint hulls and by the lemma I contains an identity modulo J, say i. If $x \in k(U)$ it follows that $ix - x$ belongs to *every* regular maximal ideal, and since R is semi-simple this means that $ix - x = 0$. Thus $x = ix \in I$, q.e.d.

We include the following variant largely for comparison with the later theory of regular Banach algebras.

Theorem. *Let R be a commutative semi-simple ring and let I be an ideal in R. Then I contains every element x such that $h(I) \subset$ int $h(x)$ and such that $x = ex$ for some $e \in R$.*

Proof. If C is the complement of $h(x)$ and $J = k(C)$, then the first hypothesis says that I and J have disjoint hulls. The second hypothesis implies that e is an identity modulo J, though this is by no means immediately obvious. It depends on the following lemma.

Lemma. *Two subsets of a semi-simple commutative ring R annihilate each other if and only if the union of their hulls is R.*

Proof. If x and y are elements such that $h(x) \cup h(y) = R$, then xy belongs to every regular maximal ideal and so $xy = 0$, by the semi-simplicity of R. Conversely let A and B be subsets of R such that $AB = 0$. If $h(A) \cup h(B) \neq R$, then there exist a regular maximal ideal M and elements $a \in A$ and $b \in B$ such that neither a nor b belongs to M. Thus neither a nor b is zero modulo M whereas $ab = 0$, contradicting the fact that R/M is a field.

Returning now to the theorem, the fact that $x(ey - y) = (xe - x)y = 0$ implies by the lemma that $ey - y$ belongs to every regular maximal ideal which does not contain x and hence belongs to J. Thus e is an identity modulo J, and $x \in I$ exactly as before.

20G. We conclude this section with the theorem on the persistence of the hull-kernel topology under homomorphisms.

Theorem. *If I is a proper ideal in a ring R, then a subset of R/I is an ideal in R/I if and only if it is of the form J/I where J is an ideal of R including I. J/I is regular and/or maximal in R/I if and only if J is regular and/or maximal in R. The space of regular maximal ideals in R/I thus corresponds to the hull of I in the space of regular maximal ideals of R, and the correspondence is a homeomorphism with respect to the hull-kernel topologies.*

Proof. Direct verification.

§ 21. SPECTRUM; ADVERSE

In this section we shall add to our repertoire the two very important notions of *spectrum* and *adverse*. These concepts arise naturally out of the question as to what can be meant in general by the statement: the element x assumes the value λ. This should have the same meaning as the statement that $x - \lambda e$ assumes the value 0, and should reduce to the ordinary meaning in the case of our model algebra $\mathcal{C}(S)$ of all continuous functions on a compact Hausdorff space. In this case there are two obvious algebraic formulations of the statement: first, that $x - \lambda$ belongs to some maximal ideal, and second, that $(x - \lambda)^{-1}$ fails to exist. We show below that these two properties are equivalent in any algebra, so that either can be taken as the desired generalization. We set $y = x - \lambda e$, and prove the equivalence for y in any ring with an identity.

21A. Theorem. *If R is a ring with an identity, then an element y has a right inverse if and only if y lies in no maximal right ideal. If y is in the center of R, then y^{-1} exists if and only if y lies in no maximal (two-sided) ideal.*

Proof. If y has a right inverse z and lies in a right ideal I, then $yz = e \in I$ and $I = R$. Thus y can lie in no proper right ideal. Conversely, if y does not have a right inverse, then the set $\{yz: z \in R\}$ is a proper right ideal containing y, and can be extended to a maximal right ideal containing y by **20A**. Thus the first statement of the theorem is valid. The same proof holds for the second statement, needing only the additional remark that since y now is assumed to commute with every element of R the set $\{yz: z \in R\}$ is now a two-sided ideal.

21B. If x is an element of an algebra A with an identity, then the set of all λ such that $(x - \lambda e)^{-1}$ does not exist, corresponding to the range of f in the case of the algebra $\mathcal{C}(S)$, is called the *spectrum* of x.

If A does not have an identity, we define the spectrum of an element x to be that set of complex numbers which becomes the spectrum of x in the above sense when A is enlarged by adding an identity, as in **20C**.

In this case $\lambda = 0$ must always be in the spectrum of an element x, for an element $x \in A$ *cannot* have an inverse in the extended algebra A_e ($x(y + \lambda e) = e \Rightarrow e = xy + \lambda x \in A$). Moreover the non-zero spectrum of x can still be determined within A by rephrasing the discussion of $(x - \lambda e)^{-1}$ so that e does not occur. In fact, after setting $x = \lambda y$ and factoring out λ, we can write $(y - e)^{-1}$, if it exists, in the form $u - e$, and the equation $(y - e)(u - e) = e$ becomes $y + u - yu = 0$, the desired condition. Such an element u must belong to A, for $u = yu - y \in A$. The following definition is clearly indicated.

In any ring R, if $x + y - xy = 0$, then y is said to be a *right adverse* of x, and x is a *left adverse* of y. We shall see below that, if x has both right and left adverses, then they are equal and unique, and this uniquely determined element is called the *adverse* of x.

The conclusions we reached above now take the following form.

Theorem. *If A is an algebra without an identity, then 0 is in the spectrum of every element, and a non-zero λ is in the spectrum of x if and only if x/λ does not have an adverse in A.*

21C. We return now to the proof that right and left adverses are equal.

Theorem. *If an element x has both right and left adverses, then the two are equal and unique.*

Proof. If u and v are left and right adverses of x, then the proof of their equality and uniqueness is equivalent to the usual proof that $e - u$ and $e - v$, the left and right inverses of $e - x$, are equal and unique, and can be derived from that proof by cancelling out e.

A better procedure, which can be systematically exploited, is to notice that the mapping $e - x \rightarrow x$ takes multiplication into a new operation $x \circ y = x + y - xy$ and takes e into 0. It follows at once that $x \circ y$ is associative and that $0 \circ x = x \circ 0 = x$. The desired proof now takes the following form:

$$u = u \circ 0 = u \circ (x \circ v) = (u \circ x) \circ v = 0 \circ v = v.$$

21D. If we carry out the proof of **21A** for an element $x - e$ and then cancel out e, we get the following replacement theorem.

Theorem. *In any ring R an element x has a right adverse if and only if there exists no regular maximal right ideal modulo which x is a left identity.*

Proof. If x has no right adverse, then the set $\{xy - y : y \in A\}$ is a right ideal not containing x modulo which x is clearly a left identity ($xy = y \mod I$ since $xy - y \in I$), and this ideal can be extended to a regular maximal ideal with the same property by **20B**. Conversely, if x has a right adverse x' and if x is a left identity for a right ideal I, then $x = xx' - x' \in I$. Then $y = xy - (xy - y) \in I$ for every $y \in R$ and $I = R$. Thus x cannot be a left identity modulo any proper right ideal.

If $x \in$ center (R), then the above theorem has the expected counterpart involving adverse and regular two-sided ideals. In particular we have the corollary:

Corollary. *If x is an element of the center of an algebra A and if $\lambda \neq 0$, then λ is in the spectrum of x if and only if x/λ is an identity modulo some regular maximal ideal.*

21E. Theorem. *If P is a polynomial without constant term, then the spectrum of $P(x)$ is exactly $P(\text{spectrum }(x))$.*

Proof. We assume an identity, enlarging the algebra if necessary. Given λ_0, let $\mu \prod (x - \mu_n e)$ be the factorization of $P(x) - \lambda_0 e$ into linear factors. Then $(P(x) - \lambda_0 e)^{-1}$ fails to exist if and only if $(x - \mu_n e)^{-1}$ fails to exist for at least one value of n. Since $P(\mu_n) - \lambda_0 = 0$ by the definition of the μ_n, we have shown that $\lambda_0 \in \text{spectrum } P(x)$ if and only if there exists $\mu_0 \in \text{spectrum } (x)$ such that $\lambda_0 = P(\mu_0)$, q.e.d.

§ 22. BANACH ALGEBRAS; ELEMENTARY THEORY

Except for minor modifications the results of § 22–25 are due to Gelfand [12]. The principal deviations are the provisions, such as the use of the adverse, regular ideals, etc., made to take care of the lack of an identity element. The existence of the identity will be assumed only in those contexts in which it, or an inverse, is explicitly mentioned.

In this section it is shown that the set of elements which have inverses (adverses) is open, from which it follows that a maximal ideal M is closed, and hence that the quotient A/M is a normed field. An elementary application of analytic function theory leads to the conclusion that there exists only one normed field, the complex number field, and this completes the first step in our representation program.

22A. The ordinary geometric series can be used in the ordinary way to prove the existence of $(e - x)^{-1}$ in a Banach algebra.

Theorem. *If $\| x \| < 1$, then x has an adverse and $e - x$ has an inverse, given respectively by $x' = -\sum_1^\infty x^n$ and $(e - x)^{-1} = e - x' = e + \sum_1^\infty x^n$, and both are continuous functions of x.*

Proof. If $y_n = -\sum_1^n x^i$, then $\| y_m - y_n \| = \| \sum_{m+1}^n x^i \| \leq \sum_{m+1}^n \| x \|^i < \| x \|^{m+1}/(1 - \| x \|) \to 0$ as $n \to \infty$. Thus the sequence $\{y_n\}$ is Cauchy, and its limit y is given by the infinite series $-\sum_1^\infty x^i$ in the usual sense. Then $x + y - xy = \lim (x + y_n - xy_n) = \lim x^n = 0$, and y is a right adverse of x. Similarly, or because y commutes with x, y is a left adverse and therefore the unique adverse of x. If e exists, then, of course, $e - y = (e - x)^{-1}$. Since the series is clearly uniformly convergent in the closed sphere $\| x \| \leqq r < 1$, it follows that the adverse (and inverse) is a continuous function of x in the open sphere $\| x \| < 1$.

The above geometric series for the inverse $(e - x)^{-1}$ is the classical Neuman series in case x is an integral operator.

Remark: $\| x' \| \leqq \| x \|/(1 - \| x \|)$.

22B. Theorem. *If y has an inverse, then so does $y + x$ whenever $\| x \| < a = 1/\| y^{-1} \|$ and $\| (x + y)^{-1} - y^{-1} \| \leqq \| x \|/ (a - \| x \|)a$. Thus the set of elements having inverses is open and the inverse function is continuous on this set.*

Proof. If $\| x \| < \| y^{-1} \|^{-1}$, then $\| y^{-1}x \| \leqq \| y^{-1} \| \| x \| < 1$ and $y + x = y(e + (y^{-1}x))$ has an inverse by **22A**. Also $(y + x)^{-1} - y^{-1} = ((e + y^{-1}x)^{-1} - e)y^{-1} = -(-y^{-1}x)'y^{-1}$, so that by the remark at the end of **22A** $\| (y + x)^{-1} - y^{-1} \| \leqq \| x \| \| y^{-1} \|^2/(1 - \| x \| \| y^{-1} \|) = \| x \|/(a - \| x \|)a$, where $a = 1/\| y^{-1} \|$.

22C. Theorem. *If y has an adverse, then so does $y + x$ whenever $\| x \| < a = (1 + \| y' \|)^{-1}$, and $\| (y + x)' - y' \| \leqq \| x \|/ (a - \| x \|)a$. Thus the set of elements having adverses is open and the mapping $y \to y'$ is continuous on this set.*

Proof. If $\| x \| < a = (1 + \| y' \|)^{-1}$, then $\| x - xy' \| \leqq \| x \|(1 + \| y' \|) < 1$ and $u = x - xy'$ has an adverse by **22A**. But $(y + x) \circ y' = x - xy' = u$, so that $y + x$ has $y' \circ u'$ as a right adverse. Similarly $y + x$ has a left adverse, and so a unique adverse equal to $y' \circ u'$. Finally, $(y + x)' - y' = (y' \circ u') - y' = u' - y'u'$, so that $\| (y + x)' - y' \| \leqq (1 + \| y' \|)\| u' \| \leqq (1 + \| y' \|)\| u \|/(1 - \| u \|) \leqq \| x \|(1 + \| y' \|)^2/(1 - \| x \|(1 + \| y' \|)) = \| x \|/(a - \| x \|)a$.

Remark: If we take $x = \lambda y$, then $u = \lambda(y - yy') = -\lambda y'$. Therefore $[(y + \lambda y)' - y']/\lambda = [(-\lambda y')' - y'(-\lambda y')']/\lambda = \sum_1^\infty (-1)^n \lambda^{n-1}[(y')^n - (y')^{n+1}] \to (y')^2 - y'$ as $\lambda \to 0$. This proves the analyticity of $(\lambda y)'$ as a function of λ, and will be of use later.

22D. *If u is a relative identity for a proper regular ideal I, then* $\rho(I, u) \geqq 1$. For if there exists an element $x \in I$ such that $\| u - x \| < 1$, then $u - x$ has an adverse a: $(u - x)a - a - (u - x) = 0$. Since x, xa and $ua - a$ are all in I, it follows that $u \in I$, a contradiction (see **20B**). It follows that:

Lemma. *If I is a proper regular ideal, then so is its closure $\bar{I}$. In particular, a regular maximal ideal is closed.*

Proof. $\bar{I}$ is clearly an ideal, and $\rho(I, u) \geqq 1$ implies that $\rho(\bar{I}, u) \geqq 1$ so that $\bar{I}$ is proper.

22E. Theorem. *If I is a closed ideal in a Banach algebra A, then A/I is a Banach algebra.*

Proof. We know already from **6B** that A/I is a Banach space. If X and Y are two of its cosets, then $\| XY \| = \text{glb} \{\| xy \|$: $x \in X$ and $y \in Y\} \leqq \text{glb} \{\| x \| \cdot \| y \|\} = \text{glb} \{\| x \| : x \in X\} \cdot \text{glb} \{\| y \| : y \in Y\} = \| X \| \cdot \| Y \|$. If I is regular and u is a relative identity, then the coset E containing u is the identity of A/I and $\| E \| = \text{glb} \{\| x \| : x \in E\} = \text{glb} \{\| u - y \| : y \in I\} = \rho(I, u) \geqq 1$ by **22D**. If A has an identity e, then $e \in E$ and $\| E \| \leqq \| e \| = 1$ so that in this case $\| E \| = 1$ and we are done. If A does not have an identity, then it may happen that $\| E \| > 1$. In this case, as we saw in § 18, we can renorm A/I with a smaller equivalent norm so that $\| E \| = 1$.

Corollary. *If I is a regular maximal ideal and if A is commutative, then it follows from **20D** and the above theorem that A/I is a normed field.*

22F. Theorem. *Every normed field is (isometrically isomorphic to) the field of complex numbers.*

Proof. We have to show that for any element x of the field there is a complex number λ such that $x = \lambda e$. We proceed by contradiction, supposing that $x - \lambda e$ is never zero and therefore that $(x - \lambda e)^{-1}$ exists for every λ. But if F is any linear functional over the field considered as a Banach space, then $F((x - \lambda e)^{-1})$, as a function of λ, is seen by direct calculation as in **22B** to have the derivative $F((x - \lambda e)^{-2})$, and is consequently analytic over the whole plane. Also $(x - \lambda e)^{-1} \to 0$ as $\lambda \to \infty$, for $(x - \lambda e)^{-1} = \lambda^{-1}(x/\lambda - e)^{-1}$, and $(x/\lambda - e)^{-1} \to -e$ as $\lambda \to \infty$ by **22A**. Thus $F((x - \lambda e)^{-1}) \to 0$ as $\lambda \to \infty$ and $F((x - \lambda e)^{-1}) \equiv 0$ by Liouville's theorem. It follows from **8C** that $(x - \lambda e)^{-1} = 0$, a contradiction.

Remark: The above proof has not made use of the fact that multiplication is commutative, except, of course, for polynomials in a single element x and its inverse. Thus it actually has been shown that the complex number field is the only normed division algebra.

§ 23. THE MAXIMAL IDEAL SPACE OF A COMMUTATIVE BANACH ALGEBRA

23A. We have proved in the preceding section all the ingredients of the following theorem, which is the basic theorem in Gelfand's theory.

Theorem. *If A is a commutative Banach algebra, then every regular maximal ideal of A is closed and of deficiency 1, every homomorphism of A onto the complex numbers is continuous with norm $\leqq 1$, and the correspondence $h \to M_h$ between such a (continuous) homomorphism and its kernel thus identifies the space Δ with the set $\mathfrak{M}$ of all regular maximal ideals.*

Proof. **22D–F** say explicitly that every regular maximal ideal is the kernel of a homomorphism of A onto the complex numbers. Conversely, the kernel of any such homomorphism is clearly a regular maximal ideal. The theorem then follows if we know that every such homomorphism is bounded by 1 (which is equivalent, in fact, to its kernel being closed). The most direct proof is to suppose that $|h(x)| > \|x\|$ for some x and notice that, if $\lambda = h(x)$, then $\|x/\lambda\| < 1$, $(x/\lambda)'$ exists and so $h(x/\lambda) \neq 1$, contradicting the definition of λ.

In clarification of the roles played by the various properties asserted in the above theorem, we prove the following lemma.

Lemma. *If A is a commutative normed algebra, then every regular maximal ideal of A is closed if and only if A has the property that the adverse of x exists whenever $\| x \| < 1$.*

Proof. We remark first that, if M is a closed regular maximal ideal with relative identity u, then $\rho(M, u) \geq 1$, for if there exists $x \in M$ such that $\| x - u \| = \delta < 1$, then $y = u - (u - x)^n$ is a sum of positive powers of x and so belongs to M, and $\| u - y \| = \| (u - x)^n \| \leq \delta^n$, proving that $u \in \overline{M} = M$, a contradiction. Suppose then that every regular maximal ideal is closed and that $\| x \| < 1$. Then x' must exist, for otherwise x is an identity relative to some regular maximal ideal, by **21D**, and $1 > \| x \| = \| x - 0 \| \geq \rho(x, M)$, contradicting the above remark.

Conversely, the existence of x' whenever $\| x \| < 1$ implies that every regular maximal ideal is closed, exactly as in **22D**.

If a commutative normed algebra has either, and hence both, of these equivalent properties then **22F** shows that every regular maximal ideal is of deficiency 1 and the rest of the above theorem then follows. In particular, the theorem holds for an algebra A of bounded functions which is inverse-closed (see **19D**).

23B. Let us review the facts of the Gelfand representation, replacing the underlying space Δ by the space $\mathfrak{M}$ of all regular maximal ideals. Let F_M be the homomorphism of Δ whose kernel is the regular maximal ideal M. The number $F_M(x)$ is explicitly determined as follows: if e_M is the identity of the field A/M and if $\bar{x}$ is the coset of A/M which contains x, then $F_M(x)$ is that complex number λ such that $\bar{x} = \lambda e_M$. It also follows from **22E** that $| F_M(x) | \leq \| x \|$.

If x is held fixed and M is varied, then $F_M(x)$ defines a complex-valued function $\hat{x}$ ($\hat{x}(M) = F_M(x)$) on the set $\mathfrak{M}$ of all regular maximal ideals of A. The mapping $x \to \hat{x}$ is then a norm-decreasing homomorphism of A onto an algebra $\hat{A}$ of complex-valued functions on $\mathfrak{M}$, the uniform norm $\| \hat{x} \|_\infty$ being used in $\hat{A}$. $\mathfrak{M}$ is given the weak topology which makes the functions $\hat{x} \in \hat{A}$ all continuous, and it was shown in **19B** that $\mathfrak{M}$ is then either compact or locally compact.

The function algebra $\hat{A}$ is not generally the whole of $\mathcal{C}(\mathfrak{M})$, nor even a uniformly closed subalgebra; it may or may not be dense in $\mathcal{C}(\mathfrak{M})$. It always shares with $\mathcal{C}(\mathfrak{M})$ the property that its regular maximal ideals correspond exactly to the points of $\mathfrak{M}$. And $\hat{A}$ is always closed under the application of analytic functions: that is, if $x \in A$ and f is analytic on the closure of the range of $\hat{x}$, then there exists $y \in A$ such that $\hat{y}(M) \equiv f(\hat{x}(M))$. These properties are all very important, and will be discussed in some detail later on, mostly in § 24.

The function $\hat{x}$ will be called the *Fourier transform* of the element x, and the homomorphism $x \rightarrow \hat{x}(M) = F_M(x)$ associated with the regular maximal ideal M a *character* of the algebra A. Actually the situation should be somewhat further restricted before these terms are used; for instance, in some contexts it is more proper to call $\hat{x}$ the *Laplace transform* of x. This section will be largely devoted to well-known examples illuminating this terminology. We start, however, with a simple preliminary theorem relating the spectrum of x to the values assumed by $\hat{x}$.

Theorem. *The range of $\hat{x}$ is either identical to the spectrum of x or to the spectrum of x with the value 0 omitted. If $\hat{x}$ never assumes the value 1, then $\hat{x}/(1 - \hat{x}) \in \hat{A}$. If A has an identity and $\hat{x}$ is never 0, then $1/\hat{x} \in \hat{A}$.*

Proof. If $\lambda \neq 0$, then λ is in the spectrum of x if and only if x/λ is a relative identity for some maximal ideal M, i.e., if and only if $F_M(x/\lambda) = 1$, or $\hat{x}(M) = F_M(x) = \lambda$, proving the first statement. (If $F_M(x) = 0$, then $x \in M$ and it follows that $0 \in$ spectrum (x).) If the spectrum of x does not contain 1, then x has an adverse y by **21D**, and from $x + y - xy = 0$ it follows that $\hat{y} = \hat{x}/(\hat{x} - 1) \in \hat{A}$. The last statement is similarly a translation of **21A**. Both of these facts will also follow from the general theorem on the application of analytic functions, proved in **24D**.

23C. As a first example let A be the algebra of sequences $a = \{a_n\}$ of complex numbers such that $\sum_{-\infty}^{+\infty} |a_n| < \infty$, with $\|a\|$ defined as this sum and multiplication defined as convolution, $(a * b)_n = \sum_{m=-\infty}^{\infty} a_{n-m} b_m$. This was example 3a of § 18.

The fact that A is a Banach algebra under these definitions will follow from later general theory (of group algebras) and can also be easily verified directly. A has an identity e where $e_0 = 1$ and $e_n = 0$ if $n \neq 0$. Let g be the element of A such that $g_1 = 1$ and $g_n = 0$ otherwise. Then g has an inverse ($g_n^{-1} = 0$ unless $n = -1$ and $= 1$ if $n = -1$) and A is simply an algebra of infinite series $a = \sum_{-\infty}^{\infty} a_n g^n$ in powers of g, under the ordinary, formal multiplication of series. Let M be any maximal ideal of A and let $\lambda = F_M(g)$. Then $| \lambda | \leq \| g \| = 1$. But $F_M(g^{-1}) = \lambda^{-1}$ and similarly $| \lambda^{-1} | \leq 1$. Thus $| \lambda | = 1$ and $\lambda = e^{i\theta_M}$ for some $\theta_M \in (-\pi, \pi]$. Then $F_M(g^n) = e^{in\theta_M}$ and $F_M(a) = \sum_{-\infty}^{\infty} a_n e^{in\theta_M}$ for any $a \in A$. Conversely any $\theta \in (-\pi, \pi]$ defines a homomorphism F of A onto the complex numbers, $F(a) = \sum_{-\infty}^{\infty} a_n e^{in\theta}$, and the kernel of F is, of course, a maximal ideal of A. Thus the space M of maximal ideals can be identified with the interval $(-\pi, \pi]$ and the transforms $\hat{a}$ are simply the continuous functions on $(-\pi, \pi]$ having absolutely convergent Fourier series. If such a function $\hat{a}$ is never zero, then, by the above theorem, its reciprocal $1/\hat{a}$ is also a function with an absolutely convergent Fourier series. This is a well-known result of Wiener.

Notice that the weak topology on $(-\pi, \pi]$ considered as a set of maximal ideals is identical to the usual topology (with π and $-\pi$ identified), for $\hat{g}(\theta) = e^{i\theta}$ is continuous in the usual topology and separates points so that **5G** can be applied.

Now let A_0 be the subset of A consisting of those sequences a whose terms a_n are all zero for negative n. A_0 is easily seen to be closed under convolution and forms a closed subalgebra of A. A_0 contains the generator g but not its inverse g^{-1}, and the above argument shows that the homomorphisms of A_0 onto the complex numbers (and hence the maximal ideals of A_0) are defined by the complex numbers in the closed unit circle $| z | \leq 1$, $F_z(a) = \sum_1^{\infty} a_n z^n$. The maximal ideal space of A_0 is thus identified with the closed unit circle (the spectrum of the generator g) and the function algebra $\hat{A}_0$ is simply the algebra of analytic functions on $| z | < 1$ whose Taylor's series converge absolutely on $| z | \leq 1$. Since $g(z) \equiv z$ separates points, it follows from **5G** as in the above example that the weak topology on $| z | \leq 1$ is identical with the usual topology.

The theorem corresponding to the Wiener theorem mentioned above is:

Theorem. *If $f(z)$ is an analytic function on $|z| < 1$ whose Taylor's series converges absolutely on $|z| \leq 1$, and if f is not zero on $|z| \leq 1$, then the Taylor's series of $1/f$ is absolutely convergent on $|z| \leq 1$.*

23D. As a third example let $A = L^1(-\infty, \infty)$ with $\| f \| = \|f\|_1$ and multiplication again as convolution $(f * g)(x) = \int_{-\infty}^{\infty} f(x - y)g(y)\, dy$ (Ex. 3b of § 18). If F is any homomorphism of A onto the complex numbers such that $| F(f) | \leq \| f \|_1$, then, since F is in particular a linear functional in $(L^1)^*$, there exists $\alpha \in L^\infty$ with $\| \alpha \|_\infty \leq 1$ such that $F(f) = \int_{-\infty}^{\infty} f(x)\overline{\alpha(x)}\, dx$. Then

$$\int_{-\infty}^{\infty}\int_{-\infty}^{\infty} f(x)g(y)\overline{\alpha(x + y)}\, dx\, dy = \int_{-\infty}^{\infty}\int_{-\infty}^{\infty} f(x - y)g(y)\overline{\alpha(x)}\, dx\, dy$$

$$= F(f * g) = F(f) \cdot F(g)$$

$$= \int_{-\infty}^{\infty} f(x)\overline{\alpha(x)}\, dx \int_{-\infty}^{\infty} g(y)\overline{\alpha(y)}\, dy$$

$$= \int_{-\infty}^{\infty}\int_{-\infty}^{\infty} f(x)g(y)\overline{\alpha(x)}\ \overline{\alpha(y)}\, dx\, dy$$

Thus $\alpha(x + y) = \alpha(x)\alpha(y)$ almost everywhere, and if we accept for now the fact (which will follow from later theory) that α can be assumed to be continuous, then this equation holds for all x and y. But the only continuous solution of this functional equation is of the form e^{ax}, and since $| e^{ax} | \leq 1$, a is of the form iy and $\alpha(x) = e^{iyx}$. The above argument can be reversed to show that every function $\alpha(x) = e^{iyx}$ defines a homomorphism. Thus the regular maximal ideals of A are in one-to-one correspondence with the points $y \in (-\infty, \infty)$ and $\hat{f}$ is the ordinary Fourier transform, $\hat{f}(y) = \int_{-\infty}^{\infty} f(x)e^{-iyx}\, dx$. These functions are easily seen to be continuous in the ordinary topology of $(-\infty, \infty)$ and it follows as usual from **5G** that the weak topology and ordinary topology coincide.

Now let $\omega(x)$ be a non-negative weight function on $(-\infty, \infty)$ such that $\omega(x+y) \leqq \omega(x)\omega(y)$ for all x and y, and let A be the subset of functions of $L^1(-\infty, \infty)$ such that $\int_{-\infty}^{\infty} |f(x)| \omega(x)\, dx < \infty$, with this integral as $\|f\|$ and multiplication taken to be convolution. The closure of A under convolution is guaranteed by the inequality $\omega(x+y) \leqq \omega(x)\omega(y)$, as the reader can easily check. If F is a norm-decreasing homomorphism of A onto the complex numbers, then, as above, there exists $\alpha \in L^\infty$ such that $\|\alpha\|_\infty \leqq 1$ and $F(f) = \int_{-\infty}^{\infty} f(x)\overline{\alpha(x)}\omega(x)\, dx$, leading this time to the functional equation $\alpha(x+y)\omega(x+y) = \alpha(x)\omega(x)\alpha(y)\omega(y)$. Therefore, $\alpha(x)\omega(x) = e^{-sx}e^{itx}$ for some s such that $e^{-sx} \leqq \omega(x)$. Conversely, any complex number $s + it$ such that $e^{-sx} \leqq \omega(x)$ for all x defines a regular maximal ideal. As an example, suppose that $\omega(x) = e^{a|x|}$ for some $a > 0$. Then the regular maximal ideals of A are in one-to-one correspondence with the strip in the complex plane defined by $|s| \leqq a$. The transform function $f(s + it) = \int_{-\infty}^{\infty} f(x)e^{-(s+it)x}\, dx$ is a bilateral Laplace transform and is analytic interior to this strip. That the usual topology is the correct maximal ideal topology follows, as in the above examples, from **5G**.

23E. We add an example which is a generalization of that in **23C**. Let A be any commutative Banach algebra with an identity and a single generator g. It may or may not happen that g^{-1} exists. In any case we assert that the maximal ideal space $\mathfrak{M}$ is in a natural one-to-one correspondence with the spectrum S of g. For $M \to \hat{g}(M)$ is a natural mapping of $\mathfrak{M}$ onto S, and since g generates A the equality $\hat{g}(M_1) = \hat{g}(M_2)$ implies $\hat{x}(M_1) = \hat{x}(M_2)$ for every $x \in A$ and hence $M_1 = M_2$, so that the mapping is one-to-one. As before $\hat{g}$ becomes identified with the function z and the weak topology induced by $\hat{g}$ is the natural topology of S in the complex plane. $\hat{A}$ is identified with an algebra of continuous complex-valued functions on S. If the interior of S is not empty (under this identification), then $\hat{x}$ is analytic on int (S)

for every $x \in A$, for $\hat{g}$ is identified with the complex variable z and $\hat{x}$ is therefore a uniform limit of polynomials in z.

§ 24. SOME BASIC GENERAL THEOREMS

We gather together here the basic tool theorems of the commutative theory. First comes the formula for computing $\| \hat{x} \|_\infty$ in terms of $\| x \|$, and then the theorem that conversely, in the semi-simple case, the norm topology in A is determined by the function algebra $\hat{A}$. The third main theorem is the earlier mentioned theorem that $\hat{A}$ is closed under the application of analytic functions. Finally we prove the existence of and discuss the *boundary* of the maximal ideal space $\mathfrak{M}$.

24A. The formula for the computation of $\| \hat{x} \|_\infty$ can be developed in a general non-commutative form if $\| \hat{x} \|_\infty$ is replaced by the *spectral norm* of x, $\| x \|_{sp}$, which is defined for x in any complex algebra as lub $\{ | \lambda | : \lambda \in$ spectrum $x \}$. The commutative formula follows from the equality $\| \hat{x} \|_\infty = \| x \|_{sp}$ proved in **23B**.

Theorem. *In any Banach algebra* $\| x \|_{sp} = \lim_{n \to \infty} \| x^n \|^{1/n}$.

Proof. We observe first that $\mu \in$ spectrum $y \Rightarrow | \mu | \leqq \| y \|$, for, if $| \mu | > \| y \|$, then $\| y/\mu \| < 1$ and y/μ has an adverse by **22A**. Thus $\| y \|_{sp} \leqq \| y \|$. Also, we know (**21E**) that $\lambda \in$ spectrum $x \Rightarrow \lambda^n \in$ spectrum x^n, so that $\| x \|_{sp} \leqq (\| x^n \|_{sp})^{1/n}$. Combining these inequalities we get $\| x \|_{sp} \leqq \| x^n \|^{1/n}$ for every n, and so $\| x \|_{sp} \leqq \underline{\lim}_{n \to \infty} \| x^n \|^{1/n}$.

It remains to be proved that $\| x \|_{sp} \geqq \overline{\lim}_{n \to \infty} \| x^n \|^{1/n}$. By the definition of spectrum and spectral norm, $(\lambda x)'$ exists for $| \lambda | < 1/\| x \|_{sp}$. If F is any functional of A^*, it follows from **22C** that $f(\lambda) = F((\lambda x)')$ is an analytic function of λ in this circle, and its Taylor's series therefore converges there. The coefficients of this series can be identified by remembering that, for small λ, $(\lambda x)' = -\sum_1^\infty (\lambda x)^n$, giving

$$f(\lambda) = F((\lambda x)') = -\sum_{n=1}^\infty F(x^n)\lambda^n.$$

It follows in particular that $| F(x^n)\lambda^n | = | F(\lambda^n x^n) | \to 0$ as $n \to \infty$ if $| \lambda | < 1/\| x \|_{sp}$. This holds for any functional

$F \in A^*$, and it follows from a basic theorem of Banach space theory proved earlier (**8F**) that there exists a bound B for the sequence of norms $\| \lambda^n x^n \|$. Thus $\| x^n \|^{1/n} \leqq B^{1/n}/| \lambda |$ and $\overline{\lim} \| x^n \|^{1/n} \leqq 1/| \lambda |$. Since λ was any number satisfying $| \lambda | < 1/\| x \|_{sp}$ we have $\overline{\lim} \| x^n \|^{1/n} \leqq \| x \|_{sp}$, as desired.

Corollary. *In any commutative Banach algebra* $\| \hat{x} \|_\infty = \lim_{n \to \infty} \| x^n \|^{1/n}$.

24B. The *radical* of a commutative algebra is the intersection of its regular maximal ideals; if the radical is zero, the algebra is said to be *semi-simple*. If A is a commutative Banach algebra, then $x \in$ radical (A) if and only if $\hat{x}(M) = 0$ for every M, that is, $\| \hat{x} \|_\infty = \lim \| x^n \|^{1/n} = 0$. Thus a necessary and sufficient condition that A be *semi-simple* is that $\hat{x} \equiv 0 \Rightarrow x = 0$; hence that the mapping $x \to \hat{x}$ of A onto $\hat{A}$ is an algebraic isomorphism. There then arises the natural question as to whether the topology of A is determined by the function algebra $\hat{A}$, or, equivalently, by the algebraic properties of A. It is not obvious that this is so, as it was for the algebra $\mathcal{C}(S)$, for now $\| \hat{x} \|_\infty$ is in general less than $\| x \|$ and the inverse mapping $\hat{x} \to x$ is not in general norm continuous. The answer is nevertheless in the affirmative and depends directly on the closed graph theorem. We prove first a more general result.

Theorem. *Let T be an algebraic homomorphism of a commutative Banach algebra A_1 onto a dense subset of a commutative Banach algebra A_2. Then:*

(1) *The adjoint transformation T^* defines a homeomorphism of the maximal ideal space $\mathfrak{M}_2$ of A_2 onto a closed subset of the maximal ideal space $\mathfrak{M}_1$ of A_1;*

(2) *If A_2 is semi-simple T is continuous.*

Proof. Since we have not assumed T to be continuous, the adjoint T^* cannot be assumed to exist in the ordinary sense. However, if α is a homomorphism of A_2 onto the complex numbers, then $\alpha(T(x))$ is a homomorphism of A_1 into the complex numbers, and since α is automatically continuous (**23A**) and $T(A_1)$ is dense in A_2 it follows that this homomorphism is onto. We naturally designate it $T^*\alpha$. It is clear that $T^*\alpha_1 \neq T^*\alpha_2$ if

$\alpha_1 \neq \alpha_2$, so that T^* is a one-to-one mapping of Δ_2 onto a subset of Δ_1. Since $T(A_1)$ is dense in A_2, the topology of Δ_2 is the weak topology defined by the algebra of functions $(T(x))^\wedge$. But $[T(x)]^\wedge(\alpha) = \alpha(Tx) = [T^*\alpha](x) = \hat{x}(T^*\alpha)$, and since the functions $\hat{x}$ define the topology of Δ_1 the mapping T^* is a homeomorphism.

Now let β_0 be any homomorphism of Δ_1 in the closure of $T^*(\Delta_2)$. That is, given ϵ and $x_1, \cdots, x_n$, there exists $\alpha \in \Delta_2$ such that $|\beta_0(x_i) - \alpha(Tx_i)| < \epsilon$, $i = 1, \cdots, n$. This implies first that, if $T(x_1) = T(x_2)$, then $\beta_0(x_1) = \beta_0(x_2)$, so that the functional α_0 defined by $\alpha_0(T(x)) = \beta_0(x)$ is single-valued on $T(A_1)$, and second that $|\alpha_0(y)| \leq \|y\|$. Thus α_0 is a bounded homomorphism of $T(A_1)$ onto the complex numbers and can be uniquely extended to the whole of A_2. We have proved that, if β_0 belongs to the closure of $T^*(\Delta_2)$, then there exists $\alpha_0 \in \Delta_2$ such that $\beta_0(x) \equiv \alpha_0(Tx)$, i.e., $\beta_0 = T^*\alpha_0$. Thus $T^*(\Delta_2)$ is closed in Δ_1, completing the proof of (1).

If $x_n \to x$ and $T(x_n) \to y$, then $\hat{x}_n \to \hat{x}$ uniformly and $(T(x_n))^\wedge \to \hat{y}$ uniformly, and since $\hat{z}(T^*(\alpha)) = (Tz)^\wedge(\alpha)$ for all $z \in A_1$, it follows that $\hat{x}(T^*(\alpha)) \equiv \hat{y}(\alpha)$, i.e., that $(Tx)^\wedge = \hat{y}$. If A_2 is semi-simple, then $y = Tx$, and the graph of T is therefore closed. The closed graph theorem (**7G**, Corollary) then implies that T is continuous, proving (2).

Corollary. *Let A be a commutative complex algebra such that the homomorphisms of A into the complex numbers do not all vanish at any element of A. Then there is at most one norm (to within equivalence) with respect to which A is a Banach algebra.*

Proof. If there are two such norms, then A is semi-simple with respect to each and the identity mapping of A into itself is, therefore, by (2) of the theorem, continuous from either norm into the other. That is, the two norms are equivalent.

24C. This is the natural point at which to ask under what circumstances the algebra $\hat{A}$ is uniformly closed, i.e., is a Banach space under its own norm. If A is semi-simple, the answer is simple.

Theorem. *A necessary and sufficient condition that A be semi-simple and $\hat{A}$ be uniformly closed is that there exist a positive con-*

stant K *such that* $\| x \|^2 \leqq K \| x^2 \|$ *for every* $x \in A$. *The mapping* $x \to \hat{x}$ *is then a homeomorphism between the algebras* A *and* $\hat{A}$.

Proof. If A is semi-simple, the mapping $x \to \hat{x}$ is an algebraic isomorphism and, if its range $\hat{A}$ is uniformly closed, then the inverse mapping is continuous by the closed graph theorem (**7G**). Thus there exists a constant K such that $\| x \| \leqq K \| \hat{x} \|_\infty$. Then $\| x \|^2 \leqq K^2 \| \hat{x} \|_\infty^2 = K^2 \| (x^2)^\wedge \|_\infty \leqq K^2 \| x^2 \|$. Thus the condition is necessary. If, conversely, such a K exists, then

$$\| x \| \leqq K^{1\!/\!2} \| x^2 \|^{1\!/\!2} \leqq K^{1\!/\!2 + 1\!/\!4} \| x^4 \|^{1\!/\!4}$$

$$\leqq \cdots \leqq K^{1\!/\!2 + \cdots + 2^{-n}} \| x^{2^n} \|^{2^{-n}}.$$

Thus $\| x \| \leqq K \lim \| x^n \|^{1/n} = K \| \hat{x} \|_\infty$. It follows that A is semi-simple and that $\hat{A}$ is complete under the uniform norm, so that the condition is sufficient.

Corollary. *A necessary and sufficient condition that* A *be isometric to* $\hat{A}$ *is that* $\| x \|^2 = \| x^2 \|$ *for every* $x \in A$.

Proof. The case $K = 1$ above.

24D. Our third theorem states that the algebra of Fourier transforms $\hat{A}$ is closed under the application of analytic functions.

Theorem. *Let the element* $x \in A$ *be given and let* $F(z)$ *be analytic in a region* R *of the complex plane which includes the spectrum of* x *(the range of the function* $\hat{x}$, *plus* 0 *if* $\mathfrak{M}$ *is not compact). Let* Γ *be any rectifiable simple closed curve in* R *enclosing the spectrum of* x. *Then the element* $y \in A$ *defined by*

$$y = \frac{1}{2\pi i} \int_\Gamma \frac{F(\lambda)}{(\lambda e - x)} \, d\lambda$$

is such that

$$\hat{y}(M) = \frac{1}{2\pi i} \int_\Gamma \frac{F(\lambda)}{\lambda - \hat{x}(M)} \, d\lambda = F(\hat{x}(M))$$

for every regular maximal ideal M. *Thus the function algebra* $\hat{A}$ *is closed under the application of analytic functions.*

Proof. For the moment we are assuming that A has an identity e. Since Γ contains no point of the spectrum of x, the element $(\lambda e - x)^{-1}$ exists and is a continuous function of λ by

22B. The integrand $F(\lambda)(\lambda e - x)^{-1}$ is thus continuous, and hence uniformly continuous, on the compact set Γ. The classical existence proof for the Riemann-Stieltjes integral shows, without any modification, that the element $y_\Delta = \sum F(\lambda_i)(\lambda_i e - x)^{-1} \, \Delta\lambda_i$ converges in the norm of A, and the integral y is defined as its limit. Since the mapping $x \rightarrow \hat{x}$ is norm decreasing, the function $\hat{y}_\Delta(M) = \sum F(\lambda_i)(\lambda_i e - \hat{x}(M))^{-1} \, \Delta\lambda_i$ converges at least as rapidly in the uniform norm to $\hat{y}(M)$. But the limit on the right, for each M, is the ordinary complex-valued Riemann-Stieltjes integral, and the theorem is proved.

If $\hat{A}$ is taken as the algebra of functions with absolutely convergent Fourier series, then the present theorem is due to Wiener, and generalizes the result of Wiener on the existence of reciprocals mentioned at the end of **23C**.

We now rewrite the above formula in a form using the adverse. First, $(\lambda e - x)^{-1} = \lambda^{-1}(e - x/\lambda)^{-1} = \lambda^{-1}(e - (x/\lambda)') = \lambda^{-1}e - \lambda^{-1}(x/\lambda)'$. Thus

$$y = \left[\frac{1}{2\pi i} \int_\Gamma \frac{F(\lambda)}{\lambda} \, d\lambda \right] e - \frac{1}{2\pi i} \int_\Gamma \frac{F(\lambda)}{\lambda} \left(\frac{x}{\lambda} \right)' d\lambda.$$

If A does not have an identity, we cannot write the first term, and we therefore define y in the general case by:

$$y = - \frac{1}{2\pi i} \int_\Gamma \frac{F(\lambda)}{\lambda} \left(\frac{x}{\lambda} \right)' d\lambda.$$

The existence proof for y is the same as in the above case, except that the inverse is replaced by the adverse. Remembering that $x'^{\wedge}(M) = \hat{x}(M)/(\hat{x}(M) - 1) = 1 + 1/(\hat{x}(M) - 1)$, we see that

$$\hat{y}(M) = - \frac{1}{2\pi i} \int_\Gamma \frac{F(\lambda)}{\lambda} \, d\lambda + \frac{1}{2\pi i} \int_\Gamma \frac{F(\lambda)}{\lambda - \hat{x}(M)} \, d\lambda.$$

The first integral will drop out, giving the desired formula $\hat{y}(M) = F(\hat{x}(M))$, if either (a) $F(0) = 0$, or (b) Γ does not enclose $z = 0$. The latter cannot happen if $\mathfrak{M}$ is not compact, for then the spectrum of x automatically contains 0 ($\hat{x}$ being zero at infinity). If however $\mathfrak{M}$ is compact and if there exists an element x such that $\hat{x}$ never vanishes, then (b) can be applied. If we take $F \equiv 1$ in this case we get $\hat{y} \equiv 1$, proving that $\hat{A}$ has an identity, and therefore, if A is semi-simple, that A has an identity.

Corollary. *If A is semi-simple and $\mathfrak{M}$ is compact, and if there exists an element x such that $\hat{x}$ does not vanish, then A has an identity.*

Remark: The curve Γ may be composed of several Jordan curves in case the spectrum of x is disconnected. We must therefore be more precise in the specification of R: it shall be taken to be a region whose complement is connected, and whose components are therefore simply connected.

24E. Our last theorem, on the notion of boundary, is due to Silov [17].

Theorem. *Let A be an algebra of continuous complex-valued functions vanishing at infinity on a locally compact space S, and suppose that A separates the points of S. Then there exists a uniquely determined closed subset $F \subset S$, called the boundary of S with respect to A, characterized as being the smallest closed set on which all the functions $|f|, f \in A$, assume their maxima.*

Proof. The assumption that A separates points is intended to include the fact that each point is separated from infinity, i.e., that the functions of A do not all vanish at any point of S. We have earlier proved in these circumstances that the weak topology induced in S by the functions of A is identical to the given topology (**5G**).

We come now to the definition of the boundary F. Let $\mathfrak{F}$ be the family of closed subsets $F \subset S$ such that each function $|f|$, $f \in A$, assumes its maximum on F. Let $\mathfrak{F}_0$ be a maximal linearly ordered subfamily of $\mathfrak{F}$ and let $F_0 = \bigcap \{F : F \in \mathfrak{F}_0\}$. Then $F_0 \in \mathfrak{F}$, for, given $f \in A$ and not identically zero, the set where $|f|$ assumes its maximum is a compact set intersecting every $F \in \mathfrak{F}_0$ and therefore intersecting F_0. F_0 is thus a lower bound for $\mathfrak{F}_0$ and therefore a minimal element of $\mathfrak{F}$.

We show that F_0 is unique by showing that any other minimal element F_1 is a subset of F_0. Suppose otherwise. Then there exists a point $p_1 \in F_1 - F_0$ and a neighborhood N of p_1 not meeting F_0. N is defined as the set $\{p : |f_i(p) - f_i(p_1)| < \epsilon,$

$i = 1, \cdots, n\}$ for some $\epsilon > 0$ and some finite set $f_1, \cdots, f_n$ of elements of A. We can suppose that $\max | f_i - c_i | \leqq 1$, where $c_i = f_i(p_1)$. Since F_1 is minimal, there exists $f_0 \in A$ such that $| f_0 |$ does not attain its maximum on $F_1 - N$. We can suppose that $\max | f_0 | = 1$ and that $| f_0 | < \epsilon$ on $F_1 - N$ (replacing f_0 by a sufficiently high power $f_0{}^n$). Then $| f_i f_0 - c_i f_0 | < \epsilon$ on the whole of F_1 and hence everywhere. Therefore if $p_0 \in F_0$ is chosen so that $f_0(p_0) = 1$, it follows that $| f_i(p_0) - c_i | < \epsilon, i = 1,$ $\cdots, n$, and so $p_0 \in N$. Thus $p_0 \in F_0 \cap N$, contradicting the fact that $N \cap F_0 = \varnothing$. Therefore $F_1 \subset F_0$ and so $F_1 = F_0$; that is, F_0 is the only minimal element of $\mathfrak{F}$.

Remark: The reason for calling this minimal set the *boundary* of S can be seen by considering the second example in **23C**. The function algebra $\hat{A}_0$ was the set of all analytic functions in $| z | < 1$ whose Taylor's series converge absolutely in $| z | \leqq 1$ and the maximal ideal space $\mathfrak{M}$ is the closed circle $| z | \leqq 1$. The ordinary maximum principle of function theory implies that the boundary of $\mathfrak{M}$ with respect to $\hat{A}_0$ is the ordinary boundary $| z | = 1$.

Corollary. *If A is a regular function algebra, then $F = S$.*

Proof. If $F \neq S$ and $p \in S - F$, then there exists by the condition of regularity (**19F**) a function $f \in A$ such that $f = 0$ on F and $f(p) \neq 0$. This contradicts the definition of the boundary, so that F must be the whole of S.

Corollary. *If A is a self-adjoint function algebra, then $F = S$.*

The proof is given later, in **26B**.

Chapter V

SOME SPECIAL BANACH ALGEBRAS

In this chapter we shall develop the theory of certain classes of Banach algebras which we meet in the study of the group algebras of locally compact Abelian groups and compact groups. § 25 treats regular commutative Banach algebras, which are the natural setting for the Wiener Tauberian theorem and its generalizations. In § 26 we study Banach algebras with an involution, which are the proper domain for the study of positive definiteness. Finally, in § 27 we develop the theory of the H^*-algebras of Ambrose, which include as a special case the L^2-group algebras of compact groups.

§ 25. REGULAR COMMUTATIVE BANACH ALGEBRAS

This section is devoted to a partial discussion of the ideal theory of a commutative Banach algebra associated with the Wiener Tauberian theorem and its generalizations. The general problem which is posed is this: given a commutative Banach algebra A with the property that every (weakly) closed set of regular maximal ideals is the hull of its kernel, when is it true that a closed ideal in A is the kernel of its hull? The Wiener Tauberian theorem says that it is true for ideals with zero hull, provided the algebra satisfies a certain auxiliary condition. Questions of this kind are difficult and go deep, and the general situation is only incompletely understood.

25A. A Banach algebra A is said to be regular if it is commutative and its Gelfand representation $\hat{A}$ is a regular function algebra. This means by definition that the weak topology for $\mathfrak{M}$ defined by $\hat{A}$ is the same as its hull-kernel topology, and (by **19F**) is equivalent to the existence, for every (weakly) closed set $C \subset \mathfrak{M}$ and for every point $M_0 \not\in C$, of a function $f \in \hat{A}$ such that $f \equiv 0$ on C and $f(M_0) \neq 0$. We show in the lemma below that the representation algebra of a regular Banach algebra possesses "local identities," and the discussion in the rest of the section holds for any regular, adverse-closed function algebra which has this property of possessing "local identities."

Lemma. *If A is a regular Banach algebra and M_0 is a regular maximal ideal of A, then there exists $x \in A$ such that $\hat{x} \equiv 1$ in some neighborhood of M_0.*

Proof. We choose $x \in A$ such that $\hat{x}(M_0) \neq 0$, and a compact neighborhood C of M_0 on which $\hat{x}$ never vanishes. Then C is hull-kernel closed by the hypothesis that A is regular, and we know that the functions of $\hat{A}$ confined to C form the representation algebra of $A/k(C)$. Since this algebra contains a function $(\hat{x})$ bounded away from 0, it follows from the analytic function theorem **24D** that it contains the constant function 1. That is, $\hat{A}$ contains a function $\hat{x}_0$ which is identically 1 on C, q.e.d.

25B. From now on A will be any adverse-closed algebra of continuous functions vanishing at infinity on a locally compact Hausdorff space S, such that $S = \Delta$, and A is regular and has the property of the above lemma.

Lemma. *If C is a compact subset of S, then there exists $f \in A$ such that f is identically 1 on C.*

Proof. If $f_1 = 1$ on B_1 and $f_2 = 1$ on B_2, then obviously $f_1 + f_2 - f_1 f_2 = 1$ on $B_1 \cup B_2$. This step is analogous to the well-known algebraic device for enlarging idempotents. By the compactness of C and the previous lemma there exist a finite number of open sets B_i covering C and functions f_i such that $f_i = 1$ on B_i. These combine to give a function $f = 1$ on $\bigcup B_i$ by repeating the above step a finite number of times.

Corollary. *If A is a semi-simple regular Banach algebra and the maximal ideal space of A is compact, then A has an identity.*

Proof. Since $\mathfrak{M}$ is compact, the lemma implies that the constant 1 belongs to the representation algebra $\hat{A}$; thus $\hat{A}$ has an identity. The semi-simplicity of A means by definition that the natural homomorphism $x \rightarrow \hat{x}$ of A onto $\hat{A}$ is an isomorphism; therefore A has an identity.

25C. Lemma. *If F is a closed subset of S and C is a compact set disjoint from F, then there exists $f \in A$ such that $f = 0$ on F and $f = 1$ on C. In fact, any ideal whose hull is F contains such an f.*

Proof. By virtue of the preceding lemma this is an exact replica of the first lemma of **20F**. However, in view of the importance of the result we shall present here a slightly different proof. Let A_C be the function algebra consisting of the functions of A confined to C; A_C is (isomorphic to) the quotient algebra $A/k(C)$. A_C has an identity by the preceding lemma, and since C is hull-kernel closed the maximal ideals of A_C correspond exactly to the points of C. Now let I be any ideal whose hull is F and let I_C be the ideal in A_C consisting of the functions of I confined to C. The functions of I do not all vanish at any point of C, for any such point would belong to the hull of I, which is F. Thus I_C is not included in any maximal ideal of A_C, and therefore $I_C = A_C$. In particular I_C contains the identity of A_C; that is, I contains a function which is identically 1 on C.

25D. Theorem. *If F is any closed subset of S, then the functions of A with compact carriers disjoint from F form an ideal $j(F)$ whose hull is F and which is included in every other ideal whose hull is F.*

Proof. The carrier of a function f is the closure of the set where $f \neq 0$. The set of functions with compact carriers disjoint from F clearly form an ideal $j(F)$ whose hull includes F. If $p \not\subset F$, then p has a neighborhood N whose closure is compact and disjoint from F. By the fundamental condition for regularity there exists $f \in A$ such that $f(p) \neq 0$ and $f = 0$ on N'. Thus f has a compact carrier $(\subset \overline{N})$ disjoint from F and $f \in j(F)$, proving that $p \not\subset h(j(F))$. Thus $p \not\subset F$ implies that $p \not\subset h(j(F))$ and $h(j(F)) = F$.

Now if I is any ideal whose hull is F and f is any function of $j(F)$, with compact carrier C disjoint from F, then by the preceding lemma I contains a function e identically one on C, so that $f = fe \in I$. Thus $j(F) \subset I$, proving the theorem.

As a corollary of this theorem we can deduce the Wiener Tauberian theorem, but in a disguise which the reader may find perfect. Its relation to the ordinary form of the Wiener theorem will be discussed in **37A**.

Corollary. *Let A be a regular semi-simple Banach algebra with the property that the set of elements x such that $\hat{x}$ has compact support is dense in A. Then every proper closed ideal is included in a regular maximal ideal.*

Proof. Let I be a closed ideal and suppose that I is included in no regular maximal ideal. We must show that $I = A$. But the hull of I is empty and therefore I includes the ideal of all elements $x \in A$ such that $\hat{x}$ has compact support. Since the latter ideal is dense in A by hypothesis, we have $I = A$ as desired.

25E. A function f is said to belong locally to an ideal I at a point p if there exists $g \in I$ such that $g = f$ in a neighborhood of p. If p is the point at infinity, this means that $g = f$ outside of some compact set.

Theorem. *If f belongs locally to an ideal I at all points of S and at the point at infinity, then $f \in I$.*

Proof. In view of the assumption on the point at infinity, we may as well suppose that S is compact and that A includes the constant functions. Then there exists a finite family of open sets U_i covering S and functions $f_i \in I$ such that $f = f_i$ on U_i. We can find open sets V_i covering S such that $\overline{V}_i \subset U_i$, and the theorem then follows from the lemma below.

Lemma. *If $f_i \in I$ and $f = f_i$ on U_i, $i = 1, 2$, and if C is a compact subset of U_2, then there exists $g \in I$ such that $f = g$ on $U_1 \cup C$.*

Proof. Let $e \in A$ be such that $e = 1$ on C and $e = 0$ on U_2'. If $g = f_2 e + f_1(1 - e)$, then $g = f_2 = f$ on C, $g = f_1 = f$ on

$U_1 - U_2$ and $g = fe + f(1 - e) = f$ on $U_1 \cap U_2$. These equations add to the fact that $g = f$ on $U_1 \cup C$, as asserted.

In applying this theorem the following lemma is useful.

Lemma. *An element f always belongs locally to an ideal I at every point not in hull (I) and at every point in the interior of hull (f).*

Proof. If $p \not\subset$ hull (I) then there exists by **25C** a function $e \in I$ such that $e = 1$ in a neighborhood $N(p)$. Then $ef \in I$ and $f = ef$ in $N(p)$, so that f belongs locally to I at p. The other assertion of the lemma follows from the fact that I contains 0.

25F. The above theorem leads to the strongest known theorem guaranteeing that an element x belong to an ideal I in a commutative Banach algebra A. We say that the algebra A satisfies the condition D (a modification of a condition given by Ditkin) if, given $x \in M \in \mathfrak{M}$, there exists a sequence $x_n \in A$ such that $\hat{x}_n = 0$ in a neighborhood V_n of M and $xx_n \to x$. If $\mathfrak{M}$ is not compact, the condition must also be satisfied for the point at infinity.

Theorem. *Let A be a regular semi-simple Banach algebra satisfying the condition D and let I be a closed ideal of A. Then I contains every element x in $k(h(I))$ such that the intersection of the boundary of hull (x) with hull (I) includes no non-zero perfect set.*

Proof. We prove that the set of points at which $\hat{x}$ does not belong locally to $\hat{I}$ is perfect (in the one point compactification of $\mathfrak{M}$). It is clearly closed. Suppose that M_0 is an isolated point and that U is a neighborhood of M_0 such that $\hat{x}$ belongs locally to $\hat{I}$ at every point of $\overline{U}$ except M_0. There exists by the condition D a sequence y_n such that $y_n x \to x$ and such that each function $\hat{y}_n$ is zero in some neighborhood of M_0. Let e be such that $\hat{e} = 0$ in U' and $\hat{e} = 1$ in a smaller neighborhood V of M_0. Then $\hat{y}_n \hat{x} \hat{e}$ belongs locally to $\hat{I}$ at every point of $\mathfrak{M}_\infty$ and therefore is an element of $\hat{I}$ by the preceding theorem. Since I is closed and $y_n x \to x$, it follows that $xe \in I$ and hence that $\hat{x}$ belongs to $\hat{I}$ at M_0 (since $xe = x$ in V). Thus the set of points at which $\hat{x}$ does not belong locally to $\hat{I}$ is perfect. Since it is included in both hull (I) and the boundary of hull (x) by the lemma in **25E**

and the assumption that $h(I) \subset h(x)$, it must be zero by hypothesis. Thus $\hat{x}$ belongs locally to I at all points, and $x \in I$ by **25E**.

We shall see in **37C** that the group algebra of a locally compact Abelian group satisfies condition D, and this gives us the strongest known theorem of Tauberian type in the general group setting.

§ 26. BANACH ALGEBRAS WITH INVOLUTIONS

We remind the reader that a mapping $x \to x^*$ defined on an algebra A is an involution if it has at least the first four of the following properties:

$$(1) \qquad\qquad x^{**} = x$$

$$(2) \qquad\qquad (x + y)^* = x^* + y^*$$

$$(3) \qquad\qquad (\lambda x)^* = \bar{\lambda} x^*$$

$$(4) \qquad\qquad (xy)^* = y^* x^*$$

$$(5) \qquad\qquad \| xx^* \| = \| x \|^2$$

(6) $-xx^*$ has an adverse ($e + xx^*$ has an inverse) for every x.

Many important Banach algebras have involutions. For instance all the examples of § 18 possess natural involutions except for 2a and 4. In the algebras of functions 1 and 5 the involution is defined by $f^* = \bar{f}$ and the properties (1) to (6) above can all be immediately verified. In the algebra 2b of bounded operators on a Hilbert space, A^* is the adjoint of A, and we have already seen in **11B** that properties (1) to (6) hold. Group algebras will be discussed in great detail later.

The existence of an involution is indispensable for much of the standard theory of harmonic analysis, including the whole theory of positive definiteness. We begin this section by investigating the elementary implications of the presence of an involution, and then prove a representation theorem for self-adjoint Banach algebras, the spectral theorem for a bounded self-adjoint operator, the Bochner theorem on positive definite functionals, and a general Plancherel theorem.

26A. Our systematic discussion in the early letters of this section gets perhaps a trifle technical and we shall try to ease matters somewhat by proving ahead of time and out of context one of the simplest and most important theorems. The reader will then be able to omit **26B** to **26E** if he wishes and go on immediately to the numbers having more classical content.

Theorem. *If A is a commutative Banach algebra with an identity and with an involution satisfying* (1)–(5), *then its transform algebra $\hat{A}$ is the algebra $\mathbb{C}(\mathfrak{M})$ of all continuous complex-valued functions on its maximal ideal space $\mathfrak{M}$ and the mapping $x \to x^*$ is an isometry of A onto $\mathbb{C}(\mathfrak{M})$.*

Proof. We first prove that, if x is self-adjoint ($x = x^*$), then $\hat{x}$ is real-valued. Otherwise $\hat{x}$ assumes a complex value $a + bi$ ($b \neq 0$), and, if $y = x + iBe$, then y assumes the value $a + i$ ($b + B$). Remembering that $y^* = x - iBe$ we see that

$$a^2 + b^2 + 2bB + B^2 \leq \| \hat{y} \|_\infty^2 \leq \| y \|^2 = \| yy^* \| = \| x^2 + B^2 e \|$$

$$\leq \| x^2 \| + B^2$$

which is a contradiction if B is chosen so that $2bB > \| x \|^2$. This argument is due to Arens [2].

For any x the elements $x + x^*$ and $i(x - x^*)$ are self-adjoint and $2x = (x + x^*) - i[i(x - x^*)]$. Since the functions $\hat{x}$ separate points in $\mathfrak{M}$ the real-valued functions of $\hat{A}$ form a real algebra separating the points of $\mathfrak{M}$ and hence, by the Stone-Weierstrass theorem, dense in $\mathbb{C}^R(\mathfrak{M})$. Therefore $\hat{A}$ is dense in $\mathbb{C}(\mathfrak{M})$. The above expression for x in terms of self-adjoint elements also proves that $(x^*)^\wedge = (\hat{x})^-$.

Finally we prove that $\hat{A}$ is isometric to A, hence complete in the uniform norm, hence identical with $\mathbb{C}(\mathfrak{M})$. If y is self-adjoint, we have exactly as in **24C** the inequality $\| y \| \leq \| y^2 \|^{1/2} \leq \cdots \leq \| y^{2^n} \|^{2^{-n}}$, and therefore $\| y \| = \| \hat{y} \|_\infty$. In general $\| x \| = \| xx^* \|^{1/2} = \| \hat{x}(x^*)^\wedge \|_\infty^{1/2} = \| \, | \hat{x} |^2 \|_\infty^{1/2} = \| \hat{x} \|_\infty$, as asserted.

26B. A commutative Banach algebra is said to be *self-adjoint* if for every $x \in A$ there exists $y \in A$ such that $\hat{y} = \hat{x}^-$ ($\bar{a}$ being the complex conjugate of a). We note several simple consequences of this definition.

Lemma 1. *If A is a self-adjoint commutative Banach algebra, then $\hat{A}$ is dense in $\mathbb{C}(\mathfrak{M})$.*

Proof. If x and y are related as above, then $(\hat{x} + \hat{y})/2$ is the real part of $\hat{x}$ and $(\hat{x} - \hat{y})/2$ is the imaginary part. It follows that the real-valued functions of $\hat{A}$ form a real algebra which separates points of $\mathfrak{M}$ and therefore, by the Stone-Weierstrass theorem, is dense in $C^R(\mathfrak{M})$. Therefore $\hat{A}$ itself is dense in $\mathfrak{C}(\mathfrak{M})$.

Corollary. *If A is self-adjoint, then $\mathfrak{M}$ is its own boundary.*

Proof. Otherwise let M_0 be a point of $\mathfrak{M}$ not in the boundary, let f be a continuous function equal to 1 at M_0 and equal to 0 on the boundary, and let $\hat{x}$ be any function of $\hat{A}$ such that $\| f - \hat{x} \|_\infty < \frac{1}{2}$. Then $| \hat{x}(M_0) | > \frac{1}{2}$ and $| \hat{x} | < \frac{1}{2}$ on the boundary, contradicting the fact that $\hat{x}$ must assume its maximum on the boundary. Thus no such M_0 can exist and $\mathfrak{M}$ equals its boundary.

Lemma 2. *If A is self-adjoint and C is a compact subset of $\mathfrak{M}$, then there exists $x \in A$ such that $\hat{x} \geqq 0$ and $\hat{x} > 0$ on C.*

Proof. For each $M \in C$ there exists $x \in A$ such that $\hat{x}(M) \neq 0$. Then $| \hat{x} |^2 \in \hat{A}$ by the definition of self-adjointness and $| \hat{x} |^2 > 0$ on an open set containing p. It follows from the Heine-Borel theorem that a finite sum of such functions is positive on C, q.e.d. This is the same argument that was used in **19C**.

Corollary. *If A is semi-simple and self-adjoint and $\mathfrak{M}$ is compact, then A has an identity.*

Proof. There exists $x \in A$ such that $\hat{x} > 0$ on $\mathfrak{M}$ and it follows from the corollary of **24D** that A has an identity.

26C. If A is self-adjoint and semi-simple, then, given x, there exists a *unique y* such that $\hat{y} = \hat{x}^-$. If this y be denoted x^*, then the mapping $x \to x^*$ is clearly an involution on A satisfying (1)–(4). It also satisfies (6), for $-| \hat{x} |^2$ never assumes the value 1, and $-xx^*$ is therefore never an identity for a regular maximal ideal, so that $-xx^*$ has an adverse by **21D**. The converse is also true, without the hypothesis of semi-simplicity.

Theorem. *If A is a commutative Banach algebra with an involution satisfying (1)–(4) and (6), then A is self-adjoint and $x^{*\wedge} = \hat{x}^-$ for every $x \in A$.*

Proof. An element x such that $x = x^*$ is called *self-adjoint*. In proving that $x^{*\wedge} = \hat{x}^-$, it is sufficient to prove that, if x is self-adjoint, then $\hat{x}$ is real-valued, for in any case $x + x^*$ and $i(x - x^*)$ are self-adjoint elements, and, if they are known to have real transforms, then the conclusion about x follows from its expression as $[(x + x^*) - i(i(x - x^*))]/2$, together with (3).

Accordingly, let x be self-adjoint and suppose that $\hat{x}$ is not real, $\hat{x}(M) = a + bi$ for some M, with $b \neq 0$. Then some linear combination of $\hat{x}(M)$ and $(x^2)^{\wedge}(M)$ has pure imaginary part, for the number pairs (a, b) and $(a^2 - b^2, 2ab)$ are linearly independent. The actual combination is $y = [(b^2 - a^2)x + ax^2]/b(a^2 + b^2)$, giving $\hat{y}(M) = i$. Then $(-y^2)^{\wedge}(M) = 1$ and $-y^2 = -yy^*$ cannot have an adverse. This contradicts (6), and proves that $\hat{x}$ is real-valued if x is self-adjoint, q.e.d.

26D. Theorem. *If a commutative Banach algebra is semi-simple, then it follows from properties (1)–(4) alone that an involution is continuous.*

Proof. **24B** can be applied almost directly. Actually, the proof of **24B** must be modified slightly to allow a mapping T of A_1 onto a dense subalgebra of A_2 which departs slightly from being a homomorphism (to the extent that involution fails to be an isomorphism). The details will be obvious to the reader.

26E. We now raise the question, suggested by **26B**, as to when a self-adjoint algebra has the property that $\hat{A} = \mathfrak{C}(\mathfrak{M})$. Some of the argument is the same as in **24C**, but will nevertheless be repeated. If $\hat{A} = \mathfrak{C}(\mathfrak{M})$ and A is semi-simple, the continuous one-to-one linear mapping $x \to \hat{x}$ of A onto $\hat{A} = \mathfrak{C}(\mathfrak{M})$ must have a continuous inverse by the closed graph theorem. Thus there exists a constant K such that $\|x\| \leq K\|\hat{x}\|_\infty$ for every x, and, in particular, $\|x\|^2 \leq K^2\|\hat{x}\|_\infty^2 = K^2\|\hat{x}\hat{x}^-\|_\infty \leq K^2\|xx^*\|$. Conversely, this condition is sufficient to prove that $\hat{A} = \mathfrak{C}(\mathfrak{M})$ even without the assumption of self-adjointness.

Theorem. *If a commutative Banach algebra A has an involution satisfying (1)–(4) and the inequality $\|x\|^2 \leq K\|xx^*\|$, then $\|x\| \leq K\|\hat{x}\|_\infty$ for all $x \in A$, A is semi-simple and self-adjoint, and $\hat{A} = \mathfrak{C}(\mathfrak{M})$.*

Proof. If y is self-adjoint, then the assumed inequality becomes $\| y \|^2 \leq K \| y^2 \|$. Applying this inductively to the powers y^{2^n}, we get $\| y \| \leq K^{\frac{1}{2}} \| y^2 \|^{\frac{1}{2}} \leq K^{\frac{1}{2}} K^{\frac{1}{4}} \| y^4 \|^{\frac{1}{4}} \leq \cdots \leq K^{\frac{1}{2}} K^{\frac{1}{4}} \cdots K^{2^{-n}} \| y^{2^n} \|^{2^{-n}}$. Therefore $\| y \| \leq K \lim \| y^m \|^{1/m} = K \| \hat{y} \|_\infty$. The equation $x^{*\wedge}(M^*) = \overline{\hat{x}(M)}$ implies that $\| x^{*\wedge} \|_\infty = \| \hat{x} \|_\infty$. Thus $\| x \|^2 \leq K \| xx^* \| \leq K^2 \| \hat{x}x^{*\wedge} \|_\infty \leq K^2 \| \hat{x} \|_\infty^2$, and $\| x \| \leq K \| \hat{x} \|_\infty$, as asserted in the theorem. One consequence of this inequality is obviously the semi-simplicity of A. Another is that the one-to-one mapping $x \to \hat{x}$ is bicontinuous and that the algebra $\hat{A}$ is therefore uniformly closed. It will follow from **26B** that $\hat{A} = \mathcal{C}(\mathfrak{M})$ if A is self-adjoint.

The equation $x^{*\wedge}(M^*) = \overline{\hat{x}(M)}$ shows that, if F is a minimal closed set on which every function $| \hat{x} |$ assumes its maximum, then so is F^*. But the only such minimal closed set is the boundary F_0, so that $F_0 = F_0^*$. We now show that $M = M^*$ if M is a point of the boundary F_0. Otherwise we can find a neighborhood U of M such that $U \cap U^* = \varnothing$, and a function $\hat{x} \in \hat{A}$ which takes its maximum absolute value in U and nowhere else on the boundary F_0. We can suppose that max $| \hat{x} | = 1$ and that $| \hat{x} | < \epsilon$ on $F_0 - U$ (replacing $\hat{x}$ by $(\hat{x})^n$ for n sufficiently large). Then $| (x^*)^\wedge | < \epsilon$ on $F_0 - U^*$ and $| \hat{x}x^{*\wedge} | < \epsilon$ on the whole boundary F_0 and hence everywhere. Then $1 \leq \| x \|^2 \leq K^2 \| \hat{x}x^{*\wedge} \|_\infty < K^2\epsilon$, a contradiction if $\epsilon < 1/K^2$. Thus $(x^*)^\wedge = \hat{x}^-$ on the boundary. Since $\hat{A}$ is uniformly closed, we see from **26B** that the restriction of $\hat{A}$ to the boundary F_0 is the algebra $\mathcal{C}(F_0)$ of all continuous complex-valued functions vanishing at infinity on F_0. We know (**19C** and **20C**) that F_0 is the set of all regular maximal ideals of this algebra, and since the restriction of $\hat{A}$ is, by the definition of the boundary, isomorphic (and isometric) to $\hat{A}$ itself, it follows that F_0 is the set of all regular maximal ideals of $\hat{A}$, i.e., $F_0 = \mathfrak{M}$. Thus $x^{*\wedge} = x^{\wedge -}$ everywhere and A is self-adjoint.

Corollary. *If A is a commutative Banach algebra with an involution satisfying* (1)–(5), *then A is isometric and isomorphic to* $\mathcal{C}(\mathfrak{M}) = \hat{A}$.

Proof. Now $K = 1$, and the inequality $\| x \| \leq K \| \hat{x} \|_\infty$ implies that $\| x \| = \| \hat{x} \|_\infty$.

Corollary. *If A is a commutative algebra of bounded operators on a Hilbert space H, closed under the norm topology and under the adjoint operation, then A is isometric and isomorphic to the algebra $\mathfrak{C}(\mathfrak{M})$ of all continuous complex-valued functions vanishing at infinity on a certain locally compact Hausdorff space $\mathfrak{M}$.*

Proof. The above corollary and **11B**.

It should perhaps be pointed out that the theorem can be proved without using the notion of boundary by adding an identity and applying **26A**. We start, as in the first paragraph of the given proof, by showing that the norm on A is equivalent to the spectral norm. The following lemma is then the crucial step.

Lemma. *If A is a commutative Banach algebra with an involution satisfying $\| x \|^2 \leqq K\| xx^* \|$, then $\| x \|_{sp}^2 = \| xx^* \|_{sp}$.*

Proof. $\| (xx^*)^n \|^{1/n} = \| x^n x^{*n} \|^{1/n} \leqq \| x^n \|^{1/n}\| x^{*n} \|^{1/n}$. Taking limits we get $\| xx^* \|_{sp} \leqq \| x \|_{sp}\| x^* \|_{sp} = \| x \|_{sp}^2$. Conversely, $\| x^n \|^2 \leqq K\| x^n(x^n)^* \| = K\| (xx^*)^n \|$, $\| x^n \|^{2/n} \leqq K^{1/n}\| (xx^*)^n \|^{1/n}$ and, letting $n \to \infty$, $\| x \|_{sp}^2 \leqq \| xx^* \|_{sp}$.

Supposing, then, that the original algebra has been renormed with its spectral norm, we now add an identity and observe that the norm $\| x + \lambda e \| = \| x \| + | \lambda |$ satisfies the K inequality with $K = 16$. For if $\| x \| < 3| \lambda |$, then $\| x + \lambda e \|^2 = (\| x \| + | \lambda |)^2 < 16| \lambda |^2 \leqq 16\| (x + \lambda e)(x^* + \bar{\lambda}e) \|$, while, if $\| x \| \geqq 3| \lambda |$, then $\| xx^* + \lambda x^* + \bar{\lambda}x \| \geqq \| xx^* \| - \| \lambda x^* + \bar{\lambda}x \| \geqq \| x \|^2/3$ and obviously $(\| x \| + | \lambda |)^2 \leqq 16(\| x \|^2/3 + | \lambda |^2)$.

The lemma therefore implies that $\| y \|_{sp}^2 = \| yy^* \|_{sp}$ in the extended algebra and the theorem is now a direct corollary of **26A**.

26F. A representation T of an algebra A is a homomorphism $(x \to T_x)$ of A onto an algebra of linear transformations over a vector space X. If S has an involution, then X is generally taken to be a Hilbert space H and T is required to take adjoints into adjoints: $T_{x^*} = (T_x)^*$. It is then called a $*$-representation (star representation).

Lemma. *If A is a Banach algebra with a continuous involution, then every $*$-representation is necessarily continuous.*

Proof. Since T is a homomorphism, T_y has an adverse whenever y does and therefore $\| T_x \|_{sp} \leqq \| x \|_{sp}$ for every x. Now T_{x^*x} is a self-adjoint operator, and for it we know (see **26A**) that $\| T_x \|^2 = \| (T_x)^*(T_x) \| = \| T_{x^*x} \| = \| T_{x^*x} \|_{sp}$. Thus $\| T_x \|^2 = \| T_{x^*x} \|_{sp} \leqq \| x^*x \|_{sp} \leqq \| x^*x \| \leqq \| x^* \| \| x \| \leqq B \| x \|^2$, where B is a bound for the involution transformation. That is, T is bounded with bound $B^{1/2}$.

If the involution is an isometry ($\| x^* \| = \| x \|$), then $B = 1$ and $\| T \| \leqq 1$ in the above argument. The theorem of Gelfand and Neumark quoted earlier (§ 11) says simply that every C^*-algebra has an isometric $*$-representation.

In this paragraph we are principally concerned with $*$-representations of commutative, self-adjoint algebras. It follows from the above argument that in this case $\| T_x \| \leqq \| | \hat{x} |^2 \|_\infty^{1/2} = \| \hat{x} \|_\infty$ so that any such representation can be transferred to a norm-decreasing representation of the function algebra $\hat{A}$, and, since $\hat{A}$ is dense in $\mathcal{C}(\mathfrak{M})$, the representation can then be extended to a norm-decreasing representation of $\mathcal{C}(\mathfrak{M})$. We now show that T has a unique extension to the bounded Baire functions on $\mathfrak{M}$.

Theorem. *Let T be a bounded representation of the algebra $\mathcal{C}(\mathfrak{M})$ of all continuous complex-valued (or real-valued) functions vanishing at infinity on a locally compact Hausdorff space $\mathfrak{M}$ by operators on a reflexive Banach space X. Then T can be extended to a representation of the algebra $\mathcal{B}(\mathfrak{M})$ of all bounded Baire functions which vanish at infinity on $\mathfrak{M}$, and the extension is unique, subject to the condition that $L_{x,y}(f) = (T_f x, y)$ is a complex-valued bounded integral for every $x \in X$, $y \in X^*$. If S is a bounded operator on X which commutes with T_f for every $f \in \mathcal{C}(\mathfrak{M})$, then S commutes with T_f for every $f \in \mathcal{B}(\mathfrak{M})$. If X is a Hilbert space H and T is a $*$-representation, then the extended representation is a $*$-representation.*

The function $F(f, x, y) = (T_f x, y)$ defined for $f \in \mathcal{C}(\mathfrak{M})$, $x \in X$, $y \in X^*$ is trilinear and $| F(f, x, y) | \leqq \| T \| \| f \|_\infty \| x \| \| y \|$. If x and y are fixed it is a bounded integral on $\mathcal{C}(\mathfrak{M})$ and hence uniquely extensible to $\mathcal{B}(\mathfrak{M})$ with the same inequality holding. The extended functional is, moreover, linear in f, x and y. If x and f are fixed, it belongs to $X^{**} = X$; that is, there exists an element of X which we designate $T_f x$ such that $F(f, x, y)$ $= (T_f x, y)$ for all y. From the linearity of F in its three arguments and the inequality $| F(f, x, y) | \leq \| T \| \, \| f \|_\infty \| x \| \, \| y \|$ we see that T_f is linear and bounded by $\| T \| \, \| f \|_\infty$ and that the mapping $f \to T_f$ is linear and bounded by $\| T \|$. Now for $f, g \in \mathbb{C}(\mathfrak{M})$ we have

(a) $$T_{fg} = T_f T_g = T_g T_f$$

or

(b) $$F(fg, x, y) = F(f, T_g x, y) = F(f, x, (T_g)^* y).$$

Keeping g fixed this identity between three integrals in f persists when the domain is extended from $\mathbb{C}(\mathfrak{M})$ to $\mathfrak{B}(\mathfrak{M})$, proving (a) for $g \in \mathbb{C}(\mathfrak{M})$ and $f \in \mathfrak{B}(\mathfrak{M})$. Since (a) is symmetric in f and g, we have (b) for $g \in \mathfrak{B}(\mathfrak{M})$ and $f \in \mathbb{C}(\mathfrak{M})$. Extending once more we have (b) and hence (a) for all $f, g \in \mathfrak{B}(\mathfrak{M})$. Thus the extended mapping $f \to T_f$ is a representation of $\mathfrak{B}(\mathfrak{M})$.

If S commutes with T_f for every $f \in \mathbb{C}(\mathfrak{M})$, we have $(T_f S x, y)$ $= (S T_f x, y) = (T_f x, S^* y)$, i.e., $F(f, S x, y) = F(f, x, S^* y)$. This identity persists through the extension, as above, and then translates back into the fact that S commutes with T_f for every $f \in \mathfrak{B}(\mathfrak{M})$.

Finally, if X is a Hilbert space H the assumption that T is a $*$-representation is equivalent to the identity $F(\bar{f}, y, x) = \overline{F(f, x, y)}$, which again persists through the extension and translates back into the fact that the extended T is a $*$-representation. This completes the proof of the theorem.

26G. The essential content of the spectral theorem is that a bounded self-adjoint operator on a Hilbert space can be approximated in the operator norm by linear combinations of projections. This is similar to the fact that a bounded continuous function on a topological space can be approximated in the uniform norm by step functions (i.e., by linear combinations of characteristic functions). The Gelfand theory reveals these apparently unconnected statements to be, in fact, equivalent assertions, and this fact gives rise to an elegant and easy proof of the spectral theorem.

Let $\mathfrak{a}$ be a commutative algebra of bounded operators on a Hilbert space H, closed under the operation of taking the adjoint, and topologically closed under the operator norm. We may as well suppose that $\mathfrak{a}$ contains the identity, for in any case it can be added. Then $\mathfrak{a}$ is isometric and isomorphic to the algebra $\mathfrak{C}(\mathfrak{M})$ of all continuous functions on its compact maximal ideal space (**26A** and **11B**) and the inverse mapping can be uniquely extended (**26F**) to the algebra $\mathfrak{B}(\mathfrak{M})$ of all bounded Baire functions on $\mathfrak{M}$. Let A be a fixed self-adjoint operator from $\mathfrak{a}$ and $\hat{A}$ its image function on $\mathfrak{M}$. (This conflicts, momentarily, with our earlier use of the symbol $\hat{A}$.) We suppose that $-1 \leqq \hat{A} \leqq 1$, and, given ϵ, we choose a subdivision $-1 = \lambda_0 < \lambda_1 < \cdots < \lambda_n = 1$ such that $\max (\lambda_i - \lambda_{i-1}) < \epsilon$. Let $\hat{E}_\lambda$ be the characteristic function of the compact set where $\hat{A} \leqq \lambda$, and choose λ_i' from the interval $[\lambda_{i-1}, \lambda_i]$. Then

and hence

$$\| \hat{A} - \textstyle\sum_1^n \lambda_i'(\hat{E}_{\lambda_i} - \hat{E}_{\lambda_{i-1}}) \|_\infty < \epsilon$$

$$\| A - \textstyle\sum_1^n \lambda_i'(E_{\lambda_i} - E_{\lambda_{i-1}}) \| < \epsilon$$

where E_λ is the bounded self-adjoint operator determined by the Baire function $\hat{E}_\lambda$. E_λ is idempotent (since $(\hat{E}_\lambda)^2 = \hat{E}_\lambda$) and hence a projection. The approximation above could be written in the form of a Riemann-Stieltjes integral,

$$A = \int \lambda \, dE_\lambda,$$

which is the integral form of the spectral theorem.

All the standard facts connected with the spectral theorem follow from the above approach. For instance, simple real-variable approximation arguments (one of which can be based on the Stone-Weierstrass theorem) show that, for fixed λ, there exists a sequence of polynomials P_n such that $P_n(\hat{A}) \downarrow \hat{E}_\lambda$. Remembering that $(P_n(A)x, x) = \int P_n(\hat{A}) \, d\mu_{x,x}$ it follows that $P_n(A)$ converges monotonically to E in the usual weak sense: $(P_n(A)x, x) \downarrow (E_\lambda x, x)$ for every $x \in H$.

26H. We conclude § 26 with a study of positive-definiteness culminating in a general Plancherel theorem. Most of the elementary facts appeared in the early Russian literature, but the Plancherel theorem itself is a modification of that given by Godement [20].

If A is a complex algebra with an involution, a linear functional φ over A is said to be *positive* if $\varphi(xx^*) \geqq 0$ for all x. The significance of positivity is that the form $[x, y] = \varphi(xy^*)$ then has all the properties of a scalar product, except that $[x, x]$ may be zero without x being zero. The linearity of $[x, y]$ in x and its conjugate linearity in y are obvious, and the only remaining property to be checked is that $[x, y] = \overline{[y, x]}$. This follows upon expanding the left member of the inequality $\varphi((x + \lambda y)(x + \lambda y)^*) \geqq 0$, showing that $\lambda \varphi(yx^*) + \bar{\lambda} \varphi(xy^*)$ is real-valued for every complex number λ, from which it follows by an elementary argument that $\varphi(yx^*) = \overline{\varphi(xy^*)}$ as required.

It then follows (see **10B**) that the Schwarz inequality is valid:

$$| \varphi(xy^*) | \leqq \varphi(xx^*)^{\frac{1}{2}} \varphi(yy^*)^{\frac{1}{2}}.$$

If A has an identity, we can take $y = e$ in the Schwarz inequality and in the equation $\varphi(xy^*) = \overline{\varphi(yx^*)}$ and get the conditions

$$| \varphi(x) |^2 \leqq k\varphi(xx^*), \quad \varphi(x^*) = \overline{\varphi(x)}$$

where $k = \varphi(e)$. In any case, a positive functional satisfying these extra conditions will be called extendable, for reasons which the following lemma will make clear.

Lemma 1. *A necessary and sufficient condition that φ can be extended so as to remain positive when an identity is added to A is that φ be extendable in the above sense.*

Proof. The necessity is obvious from the remarks already made. Supposing then that φ satisfies the above conditions and taking $\varphi(e) = k$, we have $\varphi((x + \lambda e)(x + \lambda e)^*) = \varphi(xx^*) + 2\Re\lambda\varphi(x) + | \lambda |^2 k \geqq \varphi(xx^*) - 2| \lambda | k^{\frac{1}{2}}\varphi(xx^*)^{\frac{1}{2}} + | \lambda |^2 k = (\varphi(xx^*)^{\frac{1}{2}} - | \lambda | k^{\frac{1}{2}})^2 \geqq 0$, proving the sufficiency.

Lemma 2. *If A is a Banach algebra with an identity and a continuous involution, then every positive functional on A is continuous.*

Proof. If A is a Banach algebra with an identity and if $\| x \| < 1$, then $e - x$ has a square root which can be computed by the ordinary series expansion for $\sqrt{1 - t}$ about the origin. If A has a continuous involution and x is self-adjoint, then so is the series value for $y = \sqrt{e - x}$. Thus $\varphi(e - x) = \varphi(yy^*) \geqq 0$ and $\varphi(x) \leqq \varphi(e)$. Similarly $\varphi(-x) \leqq \varphi(e)$, and we have the conclusion that, if x is self-adjoint and $\| x \| < 1$, then $| \varphi(x) | \leqq \varphi(e)$. For a general x we have the usual expression $x = (x + x^*)/2 - i[i(x - x^*)/2]$ where $x + x^*$ and $i(x - x^*)$ are self-adjoint. If B is a bound for the continuous involution, it follows that $| \varphi(x) | \leqq \sqrt{2}\varphi(e)$ whenever $\| x \| < 2/(B + 1)$, proving that φ is continuous with $(B + 1)\varphi(e)/\sqrt{2}$ as a bound.

26I. Theorem. (Herglotz-Bochner-Weil-Raikov.) *If A is a semi-simple, self-adjoint, commutative Banach algebra, then a linear functional φ on A is positive and extendable if and only if there exists a finite positive Baire measure μ_φ on $\mathfrak{M}$ such that $\varphi(x) = \int \hat{x} \, d\mu_\varphi$ for every $x \in A$.*

Proof. If $\varphi(x) = \int \hat{x} \, d\mu$ where μ is a finite positive Baire measure on $\mathfrak{M}$, then

$$\varphi(xx^*) = \int | \hat{x} |^2 \, d\mu \geqq 0,$$

$$\varphi(x^*) = \int \hat{x}^* \, d\mu = \left(\int \hat{x} \, d\mu \right)^- = \overline{\varphi(x)},$$

$$| \varphi(x) |^2 = | \int \hat{x} \, d\mu |^2 \leqq \left(\int | \hat{x} |^2 \, d\mu \right) \left(\int 1 \, d\mu \right) = \| \mu \| \varphi(xx^*);$$

that is, φ is positive and extendable. Conversely, if φ is positive and extendable, then φ is continuous (by Lemma 2 above) and

$$| \varphi(x) |^2 \leqq k\varphi(xx^*) \leqq k^{1 + \frac{1}{2}}\varphi((xx^*)^2)^{\frac{1}{2}}$$

$$\leqq \cdots \leqq k^{1 + \cdots + 2^{-n}}\varphi((xx^*)^{2^n})^{2^{-n}}$$

$$\leqq k^{1 + \cdots + 2^{-n}}\| \varphi \|^{2^{-n}}\| (xx^*)^{2^n} \|^{2^{-n}}.$$

Letting $n \to \infty$ and taking the square root we have $| \varphi(x) | \leq k|| \hat{x} ||_\infty$. Since $\hat{A}$ is dense in $\mathcal{C}(\mathfrak{M})$, the bounded linear functional I defined on $\hat{A}$ by $I_\varphi(\hat{x}) = \varphi(x)$ can be extended in a unique way to $\mathcal{C}(\mathfrak{M})$. If $f \in \mathcal{C}(\mathfrak{M})$ and $f \geq 0$, then $f^{1/2}$ can be uniformly approximated by functions $\hat{x} \in \hat{A}$ and hence f can be uniformly approximated by functions $| \hat{x} |^2$. Since $I_\varphi(| \hat{x} |^2) = \varphi(xx^*) \geq 0$ and $I_\varphi(| \hat{x} |^2)$ approximates $I_\varphi(f)$, it follows that $I_\varphi(f) \geq 0$. That is, I_φ is a bounded integral. If μ_φ is the related measure, we have the desired identity $\varphi(x) = \int \hat{x} \, d\mu_\varphi$ for all $x \in A$.

26J. The setting for the Plancherel theorem will be a semi-simple, self-adjoint commutative Banach algebra A, and a fixed positive functional φ defined on a dense ideal $A_0 \subset A$. An element $p \in A_0$ will be said to be *positive definite* if the functional θ_p defined on A by $\theta_p(x) = \varphi(px)$ is positive and extendable. Then by the Bochner theorem there is a unique bounded integral I_p on $\mathcal{C}(\mathfrak{M})$ (finite positive Baire measure μ_p on $\mathfrak{M}$) such that

$$\varphi(px) = I_p(\hat{x}) = \int \hat{x} \, d\mu_p.$$

The set of positive definite elements is clearly closed under addition and under multiplication by positive scalars, and we now observe that it contains every element of the form xx^*, $x \in A_0$. In fact, we can see directly that θ_{xx^*} is not only positive but can also be extended so as to remain positive, for if $\theta_{xx^*}(y + \lambda e) = \varphi(xx^*y + \lambda xx^*)$, then $\theta_{xx^*}((y + \lambda e)(y + \lambda e)^*) = \varphi((xy + \lambda x)(xy + \lambda x)^*) \geq 0$.

If p and q are positive definite, then $I_q(\hat{x}\hat{p}) = \varphi(pqx) = I_p(\hat{x}\hat{q})$ for every $x \in A$ and therefore $I_q(h\hat{p}) = I_p(h\hat{q})$ for every $h \in \mathcal{C}(\mathfrak{M})$. Let S_f be the support of f, i.e., the closure of the set where $f \neq 0$. If $\hat{p}\hat{q}$ is bounded away from 0 on S_f, then $h = f/\hat{p}\hat{q} \in \mathcal{C}(\mathfrak{M})$ and $I_q(f/\hat{q}) = I_p(f/\hat{p})$. We now define the functional I on $L(\mathfrak{M})$ (the set of functions in $\mathcal{C}(\mathfrak{M})$ having compact support) by $I(f) = I_p(f/\hat{p})$ where p is any positive definite element such that $\hat{p}$ is bounded away from 0 on S_f and hence such that $f/\hat{p} \in L$. Such a p always exists, for S_f is compact if $f \in L$ and by the Heine-Borel theorem we can find p of the form $x_1x_1^* + \cdots + x_nx_n^*$ such that $\hat{p} \geq 0$ and $\hat{p} > 0$ on S_f. We saw above that $I(f)$ is independent of the particular p taken, and using the p just constructed we see that $I(f) \geqq 0$ if $f \geqq 0$. Thus I is an integral. The original identity $I_p(h\hat{q}) = I_q(h\hat{p})$ shows that the integral I_p vanishes on the closed set where $\hat{p} = 0$, and this, together with the identity $I(f) = I_p(f/\hat{p})$ for all f which vanish on this closed set, implies that $I(g\hat{p}) = I_p(g)$ for all g. Therefore $I(\hat{p}) = I_p(1) = \| I_p \|$.

Finally, $\varphi(pq^*) = I_p(\bar{\hat{q}}) = I(\hat{p}\bar{\hat{q}})$ for all positive definite elements p and q, and $p \rightarrow \hat{p}$ is therefore a unitary mapping of the subspace of H_φ generated by positive definite elements onto a subspace of $L^2(I)$. We have proved the following theorem.

Theorem. *Let A be a semi-simple, self-adjoint, commutative Banach algebra and let φ be a positive functional defined on a dense ideal $A_0 \subset A$. Then there is a unique Baire measure μ on $\mathfrak{M}$ such that $\hat{p} \in L^1(\mu)$ and $\varphi(px) = \int \hat{x}\hat{p}\, d\mu$ whenever p is positive definite with respect to φ. The mapping $p \rightarrow \hat{p}$ when confined to the subspace of H_φ generated by positive definite elements is therefore a unitary mapping of this subspace onto a subspace of $L^2(\mu)$, and is extendable to the whole of H_φ if it is known that $A_0{}^2$ is H_φ-dense in A_0.*

26K. The last two theorems have assumed, needlessly, that A is a Banach algebra. A more general investigation would start with any complex algebra A having an involution and a positive functional φ on A. The Schwarz inequality for φ implies that $\varphi(xaa^*x^*) = 0$ whenever $\varphi(xx^*) = 0$ so that the set of x such that $\varphi(xx^*) = 0$ is a right ideal I, and right multiplication by a becomes a linear operator U_a on the quotient space A/I. The condition that U_a be bounded with respect to the φ scalar product is clearly that there exist a constant k_a such that $\varphi(xaa^*x^*) \leqq k_a\varphi(xx^*)$ for every $x \in A$, and in this case U_a can be uniquely extended to a bounded operator on the Hilbert space H_φ which is the completion of A/I with respect to the φ norm. If U_a is bounded for every a, then φ is said to be *unitary;* the mapping $a \rightarrow U_a$ is then easily seen to be a *-representation of A (see [20]). This is the only new concept needed in the commutative theory to formulate more abstract Bochner and Plancherel theorems. The passage to the space Δ of complex-valued homomorphisms of A is now accomplished by noting that since the mapping $a \rightarrow U_a$ is a $*$-representation of A, the maximal ideal space $\mathfrak{M}_\varphi$ of the algebra of operators U_a is thus identifiable with a closed subset of Δ. The Bochner and Plancherel theorems are now confined to this subset $\mathfrak{M}_\varphi$, which consists only of self-adjoint ideals and therefore renders unnecessary the assumption, made above, that $x^{*\wedge} = x^{\wedge-}$ on the whole of Δ. Even when A is a Banach algebra this last point is of some interest; it is easy to prove by characteristic juggling starting with the Schwarz inequality that a continuous functional is automatically unitary, and the hypothesis that $x^{*\wedge} = x^{\wedge-}$ can therefore be dropped from both of our theorems.

§ 27.   H*-ALGEBRAS

The preceding sections of this chapter will find their applications mainly in the theory of locally compact Abelian groups. In this section we give the very special analysis which is possible in the H^*-algebras of Ambrose [1], and which includes the theory of the L^2 group algebra of a compact group as a special case. *An H*-algebra is a Banach algebra H which is also a Hilbert space under the same norm, and which has an involution satisfying (1)–(4) of § 26 and the crucial connecting property*

$$(xy, z) = (y, x^*z).$$

It is also assumed that $\| x^* \| = \| x \|$ and that $x^*x \neq 0$ if $x \neq 0$. It follows that involution is conjugate unitary $((x^*, y^*) = (y, x))$ and hence that $(xy, z) = (x, zy^*)$.

The structure which can be determined for such an algebra is as follows. H can be expressed in a unique way as the direct sum of its (mutually orthogonal) minimal closed two-sided ideals. A minimal closed two-sided ideal can be expressed (but not uniquely) as a direct sum of orthogonal minimal left (or right) ideals, each of which contains a generating idempotent. It follows that a minimal closed two-sided ideal is isomorphic to a full matrix algebra (possibly infinite dimensional) over the complex number field, and operates as such an algebra under left multiplication on any of its minimal left ideals.

27A. We begin with some lemmas about idempotents and left ideals.

Lemma 1. *Let x be a self-adjoint element of H whose norm as a left multiplication operator is 1. Then the sequence x^{2n} converges to a non-zero self-adjoint idempotent.*

Proof. Let $||| y |||$ be the operator norm of y: $||| y ||| = \text{lub}_z || yz ||/|| z ||$. Since $|| yz || \leq || y || \, || z ||$, we have $||| y ||| \leq || y ||$. By hypothesis x is a self-adjoint element such that $||| x ||| = 1$. Then $||| x^n ||| = 1$ (see **11B**) and hence $|| x^n || \geq 1$ for all n. If $m > n$ and both are even, then

$$(x^m, x^n) \leq ||| x^{m-n} |||(x^n, x^n) = (x^n, x^n)$$

$$(x^m, x^m) \leq ||| x^p |||^2 (x^{n+p}, x^{n+p}) = (x^m, x^n)$$

where $2p = m - n$. Thus $1 \leq (x^m, x^m) \leq (x^m, x^n) \leq (x^n, x^n) \leq \cdots \leq (x^2, x^2)$ and (x^m, x^n) has a limit $l \geq 1$ as $m, n \to \infty$ through even integers. Hence $\lim || x^m - x^n ||^2 = \lim (x^m - x^n, x^m - x^n) = 0$, as is seen on expanding, and x^n converges to a self-adjoint element e with $|| e || \geq 1$. Since x^{2n} converges both to e and to e^2, it follows that e is idempotent.

Corollary. *Any left ideal I contains a non-zero self-adjoint idempotent.*

Proof. If $0 \neq y \in I$, then y^*y is a non-zero self-adjoint element of I and the x of the lemma can be taken as a suitable scalar multiple of y^*y. Then $e = \lim x^{2n} = \lim x^{2n}x^2 = ex^2 \in I$, q.e.d.

27B. It is clear, conversely, that the set He of left multiples of an idempotent is a closed left ideal. An idempotent e is said to be reducible if it can be expressed as a sum $e = e_1 + e_2$ of non-zero idempotents which annihilate each other; $e_1e_2 = e_2e_1 = 0$. If e is self-adjoint, we shall require e_1 and e_2 also to be self-adjoint, and then the annihilation condition is equivalent to orthogonality: $0 = (e_1, e_2) = (e_1^2, e_2^2) = (e_1e_2, e_1e_2) \Leftrightarrow e_1e_2 = 0$. Since then $|| e ||^2 = || e_1 ||^2 + || e_2 ||^2$ and since the norm of any non-zero idempotent is at least one, we see that a self-adjoint idempotent can be reduced only a finite number of times, and hence can always be expressed as a finite sum of irreducible self-adjoint idempotents.

Lemma. *I is a minimal left ideal if and only if it is of the form $I = He$, where e is an irreducible self-adjoint idempotent.*

Proof. If $e \in I$ and $e = e_1 + e_2$ is a reduction of e, then He_1 and He_2 are orthogonal left sub-ideals of I and I is not minimal. Thus if I is minimal, then every self-adjoint idempotent in I is irreducible. Since He is a sub-ideal, it also follows that $I = He$.

Now suppose that $I = He$ where e is irreducible. If I_1 is a proper sub-ideal of I and h is a self-adjoint idempotent in I_1, then $e_1 = eh = ehe$ is a self-adjoint idempotent in I_1 which commutes with e, and $e = e_1 + e_2$ ($e_2 = e - e_1$) is a reduction of e, a contradiction. (Notice that $e_1 \neq 0$ since $he_1 = heh = h^2 = h \neq 0$, and $e_1 \neq e$ since $He_1 \subset I_1 \neq He$.) Therefore, I is minimal.

We remark that every minimal left ideal is closed (since it is of the form He).

Corollary. *H is spanned by its minimal (closed) left ideals.*

Proof. Let M be the subspace spanned by the minimal left ideals of H. If $M \neq H$, then M is a proper closed left ideal, $M^\perp$ is a non-zero closed left ideal, $M^\perp$ contains a non-zero irreducible self-adjoint idempotent, and so $M^\perp$ includes a minimal left ideal, a contradiction.

Remark: If $x \in H$ and e is a self-adjoint idempotent, then xe is the projection of x on He, for $(x - xe, he) = ((x - xe)e, h) = (0, h) = 0$ so that $(x - xe) \perp He$.

27C. Theorem. *Every minimal left ideal generates a minimal two-sided ideal. The minimal two-sided ideals are mutually orthogonal, and H is the direct sum of their closures.*

Proof. Let $I = He$ be a minimal left ideal with generating self-adjoint idempotent e, and let N be the two-sided ideal generated by I. That is, N is the subspace generated by $IH = HeH$. We first observe that $N^* = N$, for $(HeH)^* = HeH$. Now suppose that N_1 is an ideal included in N. Since $N_1 I \subset N_1 \cap I$ and I is minimal, we have either that $I \subset N_1$, in which case $N_1 = N$, or else $N_1 I = 0$. In the latter case $N_1(IH) = N_1 N = 0$ and $N_1 N_1{}^* \subset N_1 N = 0$, so that $N_1 = 0$. Thus N has no proper subideals and is a minimal ideal.

It follows that $N^2 = N$, for $N^2 \subset N$ and $N^2 = NN^* \neq 0$. Now let N_1 and N_2 be distinct minimal ideals. Then $N_1 N_2 = N_2 N_1 = 0$, for $N_1 N_2 \subset N_1 \cap N_2$ and hence either $N_1 = N_1 \cap N_2 = N_2$ or $N_1 N_2 = 0$. It follows that, if $x, y \in N_1$ and $z \in N_2$, then $(xy, z) = (y, x^*z) = 0$, so that $N_1 = N_1{}^2 \perp N_2$. Thus distinct minimal ideals are orthogonal.

Since the minimal left ideals span H and every minimal left ideal is included in a minimal two-sided ideal, it follows that H is the direct sum of the closures of its minimal two-sided ideals, i.e., the direct sum of its minimal closed two-sided ideals.

Corollary. *Every closed two-sided ideal I is the direct sum of the minimal closed two-sided ideals which are included in I.*

Proof. Every minimal two-sided ideal is either included in I or orthogonal to I, and the corollary follows from the fact that the minimal ideals generate H, and so generate I.

Remark: Since the orthogonal complement of an ideal is a closed ideal, and the orthogonal complement of a minimal ideal is a maximal ideal, it follows from the Corollary that every closed two-sided ideal is the intersection of the maximal ideals including it.

27D. We come now to the analysis of a single minimal closed two-sided ideal N.

Lemma 1. *If $I = He$ is a minimal (closed) left ideal with generating idempotent e, then eHe is isomorphic to the complex number field.*

Proof. This is a classical proof from the Wedderburn structure theory. If $0 \neq x \in He$, then $0 \neq Hx \subset He$, and therefore $Hx = He$ since He is minimal. Hence there exists $b \in H$ such that $bx = e$. If $x \in eHe$, so that $x = exe$, then $(ebe)(exe) = ebx = e$. Thus eHe is an algebra with an identity in which every non-zero element has a left inverse. Therefore, every such element has an inverse, as in elementary group theory.

Thus eHe is a normed division algebra and hence is isomorphic to the complex number field by **22F**. This means, of course, that every element of eHe is of the form λe.

Lemma 2. *The following conditions are equivalent.*

(a) $\{e_\alpha\}$ *is a maximal collection of mutually orthogonal irreducible self-adjoint idempotents in* N;

(b) $\{He_\alpha\}$ *is a collection of mutually orthogonal minimal left ideals spanning* N;

(c) $\{e_\alpha H\}$ *is a collection of mutually orthogonal minimal right ideals spanning* N.

Proof. The equivalence of (a) and (b) follows from the fact that $He_\alpha \perp He_\beta$ if and only if $e_\alpha \perp e_\beta$, and **27B**. Similarly for (a) and (c). The existence of such a maximal collection is guaranteed by Zorn's lemma.

27E. Given $\{e_\alpha\}$ and N as above, we choose a fixed e_1 and consider any other e_α. Since $He_\alpha H$ generates N, $e_1 He_\alpha He_1 \neq 0$ and hence $e_1 He_\alpha \neq 0$. If $e_{1\alpha}$ is one of its non-zero elements, then $e_{1\alpha}e_{1\alpha}{}^*$ is in $e_1 He_1$ and $e_{1\alpha}$ can be adjusted by a scalar multiple so that $e_{1\alpha}e_{1\alpha}{}^* = e_1$. Then $e_{1\alpha}{}^* \subset (e_1 He_\alpha)^* = e_\alpha He_1$, and $e_{1\alpha}{}^*e_{1\alpha}$ can be checked to be an idempotent in the field $e_\alpha He_\alpha$ and hence equal to e_α. We define $e_{\alpha 1} = e_{1\alpha}{}^*$ and $e_{\alpha\beta} = e_{\alpha 1}e_{1\beta}$. The formulas

$$e_{\alpha\beta}e_{\gamma\delta} = e_{\alpha\delta} \quad \text{if} \quad \beta = \gamma$$

$$= 0 \quad \text{if} \quad \beta \neq \gamma$$

$$e_{\alpha\beta} = e_{\beta\alpha}{}^*, \quad e_{\alpha\alpha} = e_\alpha$$

$$(e_{\alpha\beta}, e_{\gamma\delta}) = 0 \quad \text{unless} \quad \alpha = \gamma \quad \text{and} \quad \beta = \delta$$

$$(e_{\alpha\beta}, e_{\alpha\beta}) = (e_1, e_1)$$

follow immediately from the definitions. Thus $(e_{\alpha\beta}, e_{\gamma\delta}) = (e_{\alpha 1}e_{1\beta}, e_{\gamma 1}e_{1\delta}) = (e_{1\beta}e_{\delta 1}, e_{1\alpha}e_{\gamma 1}) = 0$ unless $\beta = \delta$ and $\alpha = \gamma$, in which case it equals (e_1, e_1).

The set $e_\alpha He_\beta$ is a one-dimensional space consisting of the scalar multiples of $e_{\alpha\beta}$, for, given any $x \in H$, $e_\alpha x e_{\beta\alpha} \in e_\alpha He_\alpha \Rightarrow e_\alpha x e_{\beta\alpha} = ce_\alpha$ for some $c \Rightarrow e_\alpha x e_\beta = ce_{\alpha\beta}$.

We know that ye_β is the projection of y on He_β, and $e_\alpha y$ that of y on $e_\alpha H$, for any $y \in H$. Therefore, if $x \in N$ we have $x =$

$\sum_{\beta} x e_{\beta} = \sum_{\alpha,\beta} e_{\alpha} x e_{\beta} = \sum_{\alpha\beta} c_{\alpha\beta} e_{\alpha\beta}$, proving that the set of elements $e_{\alpha\beta}$ is an orthogonal basis for N, and that $e_{\alpha} x e_{\beta}$ is the projection of x on $e_{\alpha} H e_{\beta}$. The Fourier coefficient $c_{\alpha\beta}$ can be computed in the usual way, $c_{\alpha\beta} = (x, e_{\alpha\beta})/\| e_{\alpha\beta} \|^2$. In view of the displayed equations above, it is clear that we have proved the following theorem:

Theorem. *The algebra N is isomorphic to the algebra of all complex matrices $\{c_{\alpha\beta}\}$ such that $\sum_{\alpha,\beta} | c_{\alpha\beta} |^2 < \infty$, under the correspondence $x \leftrightarrow \{c_{\alpha\beta}\}$ where $x = \sum_{\alpha,\beta} c_{\alpha\beta} e_{\alpha\beta}$ and $c_{\alpha\beta} = (x, e_{\alpha\beta})/ \| e_1 \|^2$.*

The following theorem is implicit in the above analysis.

Theorem. *Two minimal left ideals of H are (operator) isomorphic if and only if they are included in the same minimal closed two-sided ideal N.*

Proof. Let I_1 and I_2 be the two ideals, with self-adjoint generating idempotents e_1 and e_2 and suppose that I_1 and I_2 are both included in N. If I_1 and I_2 are not orthogonal, then the mapping $x \rightarrow x e_1$ maps I_2 into a non-zero subideal of I_1, and since I_1 and I_2 are minimal the mapping is an isomorphism and onto. The fact that such a mapping is an operator isomorphism is due merely to the associative law: $y(x e_1) = (yx)e_1$. If I_1 and I_2 are perpendicular then they can be extended to a maximal collection of orthogonal minimal ideals as in the above theorem and then the mapping $x \rightarrow x e_{21}$ furnishes the required isomorphism.

Conversely, if I_1 and I_2 are included in orthogonal minimal two-sided ideals N_1 and N_2, then they cannot be operator isomorphic since $N_1(I_2) = 0$, whereas $N_1(I_1) = I_1$.

27F. Theorem. *The following statements are equivalent:*
(a) *The minimal closed two-sided ideal N is finite dimensional.*
(b) *N contains an identity, as a subalgebra of H.*
(c) *N contains a non-zero central element.*

Proof. If N is finite dimensional, then the finite sum $e = \sum_{\alpha} e_{\alpha}$ is a self-adjoint idempotent such that, for every $x \in N$, $x = \sum_{\alpha} x e_{\alpha} = xe$ and $x = \sum_{\alpha} e_{\alpha} x = ex$. Thus e is an identity for N.

For any $x \in H$ we have $xe = e(xe) = (ex)e = ex$, so that e is an element of the center of H. Thus (a) $\Rightarrow$ (b) and (c). If e is an identity for N, then $e_\alpha = ee_\alpha$ is the projection of e on He_α and so $e = \sum e_\alpha$. Since the elements e_α all have the same norm, this sum must be finite and N is finite dimensional. Finally, suppose that N contains a non-zero central element x. Since $xe_\alpha = (xe_\alpha)e_\alpha = e_\alpha xe_\alpha$, the projection of x on He_α is of the form $c_\alpha e_\alpha$ for some constant c_α, and $x = \sum c_\alpha e_\alpha$. Then $c_\beta e_{\beta\delta} = xe_{\beta\delta} = e_{\beta\delta}x = c_\delta e_{\beta\delta}$, so that $c_\beta = c_\delta$ and the coefficients c_α are all equal. Thus again the sum must be finite and N is finite dimensional. Also we have seen that any central element of N is of the form ce where c is the common value of the coefficients c_α.

Now let $\{N_\alpha\}$ be the set of minimal closed two-sided ideals of H. Then every $x \in H$ has a unique expansion $x = \sum x_\alpha$, where x_α is the projection of x on N_α. It is clear that $(xy)_\alpha = x_\alpha y = x_\alpha y_\alpha$, and it follows in particular that x_α is central if x is central: $x_\alpha y = (xy)_\alpha = (yx)_\alpha = yx_\alpha$. Therefore, given α, either $x_\alpha = 0$ for every central element x or else N_α has an identity e_α. We thus have the following corollary.

Corollary. *If H_0 is the subspace of H which is the direct sum of its finite dimensional minimal ideals, then every central element x is in H_0 and has an expansion of the form $x = \sum c_\alpha e_\alpha$, where c_α is a scalar and e_α is the identity of N_α. H_0 is the closed two-sided ideal generated by the central elements of H.*

27G. If H is commutative, then every element is central and the minimal ideals N_α are all one-dimensional, each consisting of the scalar multiples of its identity e_α. The minimal ideals N_α are the orthogonal complements of the maximal ideals M_α, and the homomorphism $x \to \hat{x}(M_\alpha)$ is given by $\hat{x}(M_\alpha) = (x, e_\alpha)/\| e_\alpha \|^2$. That is, $\hat{x}(M_\alpha)$ is simply the coefficient of x in its expansion with respect to the orthogonal basis $\{e_\alpha\}$. This follows from the fact that the isomorphism between H/M_α and the complex number field is given by $H/M_\alpha \cong N_\alpha = \{ce_\alpha\}$, and $ce_\alpha \leftrightarrow c$.

The continuous function $\hat{e}_\beta$ thus has the value 1 at M_β and is otherwise zero, so that the space $\mathfrak{M}$ of maximal ideals has the discrete topology.

Chapter VI

THE HAAR INTEGRAL

In this chapter we shall prove that there exists on any locally compact group a unique left invariant integral (or measure) called the Haar integral. For the additive group R of real numbers, the group R/I of the reals modulo 1 and the Cartesian product R^n (Euclidean n-space) this is the ordinary Lebesgue integral. For any discrete group, e.g., the group I of the integers, the Haar measure attaches to each point the measure 1. In the case of an infinite Cartesian product of unit intervals (groups R/I) it is the so-called toroidal measure of Jessen. In each of these cases Haar measure is the obvious measure already associated with the given space in terms of its known structure. However, in the case of matrix groups it is not so clear what the measure is in terms of structural considerations. And in more complicated groups the reverse situation has held—the invariant measure has been used as an aid in discovering structure, rather than the structure giving rise to the measure. A general non-structural proof of the existence of an invariant measure is therefore of the utmost importance. The first such proof, and the model for all later proofs, was given by Haar in 1933 [22].

The first three sections center around the Haar integral, § 28 developing the necessary elementary topological properties of groups, § 29 giving the existence and uniqueness proofs, and § 30 treating the modular function. The group algebra and some elementary theorems on representations are discussed in § 31 and § 32, and the chapter closes with the theory of invariant measures on quotient spaces in § 33.

§ 28.   THE TOPOLOGY OF LOCALLY COMPACT GROUPS

A topological group is a group G together with a topology on G under which the group operations are continuous. Any group is a topological group under the discrete topology, and in fact is locally compact. The special groups R, R/I, R^n, $(R/I)^{\aleph}$ mentioned in the introduction above are all locally compact Abelian topological groups; the second and fourth are compact. A simple example of a non-Abelian locally compact group is the group of all linear transformations $y = ax + b$ of the straight line into itself such that $a > 0$, the topology being the ordinary topology of the Cartesian half-plane of number pairs $\langle a, b \rangle$ such that $a > 0$. More generally any group of $n \times n$ matrices which is closed as a subset of Euclidean n^2-dimensional space is a locally compact group which is usually not Abelian. Groups such as these are interesting to us because, being locally compact, they carry invariant measures by the general theorem of § 29 and therefore form natural domains for general harmonic analysis. Of course, there are large classes of topological groups which are not locally compact and which are of great importance in other fields of mathematics.

In this section we develop the minimum amount of topological material necessary for integration theory.

28A. We start with some simple immediate consequences of the definition of a topological group.

1) If a is fixed, the mapping $x \rightarrow ax$ is a homeomorphism of G onto itself, taking the part of G around the identity e into the part around a. If V is any neighborhood of e, then aV is a neighborhood of a, and if U is any neighborhood of a, then $a^{-1}U$ is a neighborhood of e. Thus G is "homogeneous" in the sense that its topology around any point is the same as around any other point. The mapping $x \rightarrow xa$ is another homeomorphism taking e into a. The inverse mapping $x \rightarrow x^{-1}$ is also a homeomorphism of G onto itself.

2) Every neighborhood U of e includes a *symmetric* neighborhood W of e, that is, one such that $W = W^{-1}$. In fact $W = U \cap U^{-1}$ has this property. Similarly if f is continuous and vanishes off W, then $g(x) = f(x) + f(x^{-1})$ is symmetric ($g(x) =$

$g(x^{-1})$) and vanishes off W, while $h(x) = \overline{f(x)} + f(x^{-1})$ is Hermitian symmetric $\overline{(h(x)} = h(x^{-1}))$.

3) Every neighborhood U of e includes a neighborhood V of e such that $V^2 = V \cdot V \subset U$, for the inverse image of the open set U under the continuous function xy is an open set in $G \times G$ containing (e, e), and hence must include a set of the form $V \times V$. Similarly, by taking V symmetric, or by considering the continuous functions xy^{-1} and $x^{-1}y$, we can find V such that $VV^{-1} \subset U$ or such that $V^{-1}V \subset U$.

We notice that $V^{-1}V \subset U$ if and only if $aV \subset U$ whenever $e \in aV$.

4) The product of two compact sets is compact. For if A and B are compact, then $A \times B$ is compact in $G \times G$, and with this set as domain, the range of the continuous function xy is AB, which is therefore compact (see **2H**).

5) If A is any subset of G, then $\bar{A} = \bigcap_V AV$, the intersection being taken over all neighborhoods V of the identity e. For if $y \in \bar{A}$ and V is given, then yV^{-1} is an open set containing y and so contains a point $x \in A$. Thus $y \in xV \subset AV$, proving that $\bar{A} \subset AV$ for every such V, and hence that $\bar{A} \subset \bigcap_V AV$. Conversely, if $y \subset \bigcap_V AV$, then yV^{-1} intersects A for every V and $y \in \bar{A}$. Thus $\bigcap_V AV = \bar{A}$.

6) If V is a symmetric compact neighborhood of e, then V^n is compact for every n and the subgroup $V^\infty = \lim_{n \to \infty} V^n$ is thus σ-compact (a countable union of compact sets). It is clearly an open set, since $y \in V^\infty$ implies $yV \subset V^\infty V = V^\infty$, and since $V^{\infty -} \subset V^\infty V = V^\infty$ it is also closed. The left cosets of V^∞ are all open-closed sets, so that G is a (perhaps uncountable) union of disjoint σ-compact open-closed sets. This remark is important because it ensures that no pathology can arise in the measure theory of a locally compact group (see **15D**).

28B. Theorem. *If $f \in L$ (the algebra of continuous functions with compact support), then f is left (and right) uniformly continuous. That is, given ϵ, there exists a neighborhood V of the identity e such that $s \in V$ implies that $|f(sx) - f(x)| < \epsilon$ for all x, or, equivalently, such that $xy^{-1} \in V$ implies that $|f(x) - f(y)| < \epsilon$ for all x and y.*

Proof. Choose a compact set C such that $f \in L_C$ (i.e., such that $f = 0$ off C) and a symmetric compact neighborhood U of the identity. By **5F** the set W of points s such that $|f(sx) - f(x)| < \epsilon$ for every $x \in UC$ is open, and it clearly contains the identity $s = e$. If $s \in U$, then $f(sx)$ and $f(x)$ both vanish outside of UC. Therefore if $s \in V = W \cap U$, then $|f(sx) - f(x)| < \epsilon$ for every x, q.e.d.

Given a function f on G we define the functions f_s and f^s by the equations $f_s(x) = f(sx)$, $f^s(x) = f(xs^{-1})$. They can be thought of as left and right translates of f. The choice of s in one definition and s^{-1} in the other is necessary if we desire the associative laws $f_{st} = (f_s)_t$, $f^{st} = (f^s)^t$. (For other reasons f_s is often defined by $f_s(x) = f(s^{-1}x)$.)

Corollary. *If $f \in L$, $1 \leqq p \leqq \infty$ and I is an integral on L, then f_s and f^s, as elements of $L^p(I)$, are continuous functions of s.*

Proof. The case $p = \infty$ is simply the uniform continuity of f, for we saw above that $\|f_s - f\|_\infty < \epsilon$ whenever $s \in V$. Using this same V and choosing C so that $f \in L_C$, let B be a bound for I (see **16C**) on the compact set $\overline{V}C$. Then $\|f_s - f\|_p = I(|f_s - f|^p)^{1/p} \leqq B^{1/p}\|f_s - f\|_\infty < B^{1/p}\epsilon$ if $s \in V$, proving the continuity of f_s in L^p. Similarly for f^s.

28C. Quotient spaces. If S is any class and $\prod$ is a partition of S into a family of disjoint subclasses (equivalence classes), then the *quotient space* $S/\prod$, or the fibering of S by $\prod$, is obtained by considering each equivalence class as a single point. If S is a topological space, there are two natural requirements that can be laid down for a related topology on $S/\prod$. The first is that the natural mapping α of S onto $S/\prod$, in which each point of S maps into the equivalence class containing it, be continuous. Thus a subset A of $S/\prod$ should not be taken as an open set unless $\alpha^{-1}(A)$ is open in S. The second requirement is that, if f is any continuous function on S which is constant on each equivalence class of $\prod$, then the related function F on $S/\prod$ defined by $F(\alpha(x)) = f(x)$ should be continuous on $S/\prod$. In order to realize this second requirement the topology for $S/\prod$ is made as strong (large) as possible subject to the first requirement: a subset A of $S/\prod$ is defined to be open if and only if $\alpha^{-1}(A)$ is an open subset of S. Then the mapping α is continuous, by definition. Moreover, if f is continuous on S and constant on the sets of $\prod$, then $\alpha^{-1}(F^{-1}(U)) = f^{-1}(U)$ is open whenever U is open. Therefore, $F^{-1}(U)$ is open whenever U is open and the related function F is continuous.

Clearly a subset B of $S/\prod$ is closed if and only if $\alpha^{-1}(B)$ is closed in S. In particular, if each equivalence class of $\prod$ is a closed subset of S, then each point of $S/\prod$ is a closed set in $S/\prod$, and $S/\prod$ is what is called a T_1-space.

These considerations apply in particular to the quotient space of left cosets of a subgroup H in a topological group G. Moreover, such fiberings G/H have further properties which do not hold in general. Thus:

Theorem. *If H is a subgroup of a topological group G, then the natural mapping α of G onto the left coset space G/H is an open mapping, that is, $\alpha(A)$ is open if A is open.*

Proof. Suppose that A is open. We have $\alpha^{-1}(\alpha A) = AH = \bigcup \{Ax : x \in H\}$, which is a union of open sets and therefore open. But then αA is open by definition of the topology in G/H, q.e.d.

Corollary 1. *If G is locally compact, then G/H is locally compact, for any subgroup H.*

Proof. If C is a compact subset of G, then $\alpha(C)$ is a compact subset of G/H (since α is a continuous mapping—see **2H**). If $x \in$ interior (C), then the above theorem implies that $\alpha(x) \in$ interior $(\alpha(C))$. Thus compact neighborhoods map into compact neighborhoods, proving the corollary.

Corollary 2. *If G is locally compact and B is a compact closed subset of G/H, then there exists a compact set $A \subset G$ such that $B = \alpha(A)$.*

Proof. Let C be a compact neighborhood of the identity in G and choose points $x_1, \cdots, x_n$ such that $B \subset \bigcup_1^n \alpha(x_i C) = \alpha(\bigcup_1^n x_i C)$. Then $\alpha^{-1}(B) \cap (\bigcup_1^n x_i C)$ is a compact subset of G, and its image under α is B since it is simply the set of points in $\bigcup_1^n x_i C$ which map into B.

28D. We now show that for the purposes of integration theory a topological group can be assumed to be a Hausdorff space.

Theorem. *A topological group G has a minimal closed normal subgroup, and hence a maximal quotient group which is a T_1-space.*

Proof. Let H be the smallest closed set containing the identity. We show that H is a subgroup. For if $y \in H$, then $e \in Hy^{-1}$ and so $H \subset Hy^{-1}$ (H being the *smallest* closed set containing e). Thus if $x, y \in H$, then $x \in Hy^{-1}$ and $xy \in H$. Since H^{-1} is closed, it follows similarly that $H \subset H^{-1}$. Thus H is closed under multiplication and taking inverses; that is, H is a subgroup.

Moreover, H is normal, for $H \subset xHx^{-1}$ for every x and consequently $x^{-1}Hx \subset H$ for every x, so that $H = xHx^{-1}$ for every x.

Since H is the smallest subgroup of G which is closed, G/H is the largest quotient group of G which has a T_1 topology.

Lemma. *A topological group which is a T_1-space is a Hausdorff space.*

Proof. If $x \neq y$, let U be the open set which is the complement of the point xy^{-1} and choose a symmetric neighborhood V of the identity so that $V^2 \subset U$. Then V does not intersect Vxy^{-1} and Vy and Vx are therefore disjoint open sets containing y and x respectively.

Every continuous function on a topological group G is constant on all the cosets of the minimal closed subgroup H. It follows that the same is true for all Baire functions. Thus the continuous functions and Baire functions on G are, essentially, just those on G/H, so that from the point of view of integration theory we may replace G by G/H, or, equivalently, we may assume that G is a Hausdorff space. For simplicity this property will be assumed for all groups from now on, although many results (for example, those of the next three numbers) will not depend on it.

§ 29. THE HAAR INTEGRAL

We shall follow Weil [48] in our proof of the existence of Haar measure and then give a new proof of its uniqueness. This procedure suffers from the defect that the axiom of choice is invoked in the existence proof, here in the guise of choosing a point in a Cartesian product space, and then is demonstrated to be theoretically unnecessary when the chosen functional is found to be unique. There are proofs which avoid this difficulty by simultaneously demonstrating existence and uniqueness (see [7] and [33]), but they are more complicated and less intuitive, and we have chosen to sacrifice the greater elegance of such a proof for the simplicity of the traditional method.

29A. If f and g are non-zero functions of L^+, then there exist positive constants c_i and points s_i, $i = 1, \cdots, n$, such that

$$f(x) \leqq \sum_1^n c_i g(s_i x).$$

For example, if m_f and m_g are the maximum values of f and g respectively, then the c_i can all be taken equal to any number greater than m_f/m_g. *The Haar covering function* $(f; g)$ *is defined as the greatest lower bound of the set of all sums* $\sum_1^n c_i$ *of coefficients of such linear combinations of translates of* g.

The number $(f; g)$ is evidently a rough measure of the size of f relative to g, and the properties listed below, which depend directly upon its definition, show that if g is fixed it behaves somewhat like an invariant integral.

(1) $(f_s; g) = (f; g).$

(2) $(f_1 + f_2; g) \leqq (f_1; g) + (f_2; g).$

(3) $(cf; g) = c(f; g)$ where $c > 0.$

(4) $f_1 \leqq f_2 \Rightarrow (f_1; g) \leqq (f_2; g).$

(5) $(f; h) \leqq (f; g)(g; h).$

(6) $(f; g) \geqq m_f/m_g.$

In order to see (6) we choose x so that $f(x) = m_f$, and then have $m_f \leqq \sum c_i g(s_i x) \leqq (\sum c_i) m_g$, so that $(\sum c_i) \geqq m_f/m_g$. (5) follows from the remark that, if $f(x) \leqq \sum c_i g(s_i x)$ and $g(x) \leqq \sum d_j h(t_j x)$, then $f(x) \leqq \sum_{i,j} c_i d_j h(t_j s_i x)$. Thus $(f; h) \leqq$ glb $\sum c_i d_j = (\text{glb} \sum c_i)(\text{glb} \sum d_j) = (f; g)(g; h)$. The other properties are obvious.

29B. In order to make $(f; g)$ a more accurate estimate of relative size, it is necessary to take g with smaller and smaller support, and then, in order that the result be an absolute estimate of the size of f, we must take a ratio. We accordingly fix $f_0 \in L^+$ ($f_0 \neq 0$) for once and for all, and define $I_\varphi(f)$ as $(f; \varphi)/(f_0; \varphi)$. It follows from (5) that

(7) $$1/(f_0; f) \leqq I_\varphi(f) \leqq (f; f_0)$$

so that $I_\varphi f$ is bounded above and below independently of φ. Also, it follows from (1), (2) and (3) that I_φ is *left invariant, subadditive and homogeneous*. We now show that, for small φ, I_φ is nearly additive.

Lemma. *Given f_1 and f_2 in L^+ and $\epsilon > 0$, there exists a neighborhood V of the identity such that*

(8) $$I_\varphi f_1 + I_\varphi f_2 \leqq I_\varphi(f_1 + f_2) + \epsilon$$

for all $\varphi \in L^+_V$.

Proof. Choose $f' \in L^+$ such that $f' = 1$ on the set where $f_1 + f_2 > 0$. Let δ and ϵ' for the moment be arbitrary and set $f = f_1 + f_2 + \delta f'$, $h_i = f_i/f$, $i = 1, 2$. It is understood that h_i is defined to be zero where $f = 0$ and it is clear that $h_i \in L^+$. Choose V such that $|h_i(x) - h_i(y)| < \epsilon'$ whenever $x^{-1}y \in V$ (by **28B**). If $\varphi \in L^+_V$ and $f(x) \leqq \sum c_j\varphi(s_jx)$, then $\varphi(s_jx) \neq 0 \Rightarrow |h_i(x) - h_i(s_j^{-1})| < \epsilon'$, and

$$f_i(x) = f(x)h_i(x) \leqq \sum_j c_j\varphi(s_jx)h_i(x)$$
$$\leqq \sum_j c_j\varphi(s_jx)[h_i(s_j^{-1}) + \epsilon']$$
$$(f_i; \varphi) \leqq \sum_j c_j[h_i(s_j^{-1}) + \epsilon']$$
$$(f_1; \varphi) + (f_2; \varphi) \leqq \sum c_j[1 + 2\epsilon']$$

Since $\sum c_j$ approximates $(f; \varphi)$, we have

$$I_\varphi f_1 + I_\varphi f_2 \leqq I_\varphi f[1 + 2\epsilon'] \leqq [I_\varphi(f_1 + f_2) + \delta I_\varphi(f')][1 + 2\epsilon'].$$

The lemma follows if δ and ϵ' are initially chosen so that $2\epsilon'(f_1 + f_2; f_0) + \delta(1 + 2\epsilon')(f'; f_0) < \epsilon$.

29C. We have seen above that I_φ becomes more nearly an invariant integral the smaller φ is taken, and it remains only to take some kind of generalized limit with respect to φ to get the desired invariant integral.

Theorem. *There exists a non-trivial non-negative left invariant integral on L.*

Proof. For every non-zero $f \in L^+$ let S_f be the closed interval $[1/(f_0; f), (f; f_0)]$, and let S be the compact Hausdorff space $\prod_f S_f$ (the Cartesian product of the spaces S_f). For each non-zero $\varphi \in L^+$ the functional I_φ is a point in S, its f coordinate (its projection on S_f) being $I_\varphi f$. For each neighborhood V of the identity in G, let C_V be the closure in S of the set $\{I_\varphi : \varphi \in L_V{}^+\}$. The compact sets C_V have the finite intersection property since $C_{V_1} \cap \cdots \cap C_{V_n} = C_{(V_1 \cap \cdots \cap V_n)}$. Let I be any point in the intersection of all the C_V. That is, given any V and given $f_1, \cdots, f_n$ and $\epsilon > 0$, there exists $\varphi \in L_V{}^+$ such that $| I(f_i) - I_\varphi(f_i) | < \epsilon$, $i = 1, \cdots, n$. It follows from this approximation and **29B** that I is non-negative, left invariant, additive and homogeneous, and that $1/(f_0; f) \leq I(f) \leq (f; f_0)$. The extension of I from L^+ to L, obtained by defining $I(f_1 - f_2)$ as $I(f_1) - I(f_2)$, is therefore a non-trivial left invariant integral, as desired.

The integral I is extended to the class of non-negative Baire functions as in Chapter 3 and the whole machinery of Lebesgue theory is available, as set forth in that chapter. We emphasize only one fact: that a locally compact group is a union of disjoint open-closed σ-compact subsets (**28A**, 6) and therefore the difficulties which can plague non σ-finite measures do not arise here. For instance the proof that $(L^1)^* = L^\infty$ as sketched in **15D** is available. (See also **13E**.)

29D. Theorem. *The above integral is unique to within a multiplicative constant.*

Proof. Let I and J be two left invariant non-negative integrals over L and let $f \in L^+$. We choose a compact set C such that $f \subset L_C{}^+$, an open set U with compact closure such that $C \subset U$ and a function $f' \in L^+$ such that $f' = 1$ on U. Given ϵ, we furthermore choose a symmetric neighborhood V of the identity such that $\| f_y - f^z \|_\infty < \epsilon$ if $y, z \in V$, and such that $(CV \cup VC) \subset U$. The latter condition guarantees that $f(xy) = f(xy)f'(x)$ and $f(yx) = f(yx)f'(x)$ if $y \in V$, and together with the former implies that $| f(xy) - f(yx) | < \epsilon f'(x)$ for all x. Let

h be any non-zero symmetric function in $L_V{}^+$. The above conditions, together with the Fubini theorem and the left invariance of I and J, imply that

$$I(h)J(f) = I_y J_x h(y)f(x) = I_y J_x h(y)f(yx)$$

$$J(h)I(f) = I_y J_x h(x)f(y) = I_y J_x h(y^{-1}x)f(y)$$

$$= J_x I_y h(x^{-1}y)f(y) = J_x I_y h(y)f(xy)$$

$$|\, I(h)J(f) - J(h)I(f)\,| \leqq I_y J_x (h(y)|\, f(yx) - f(xy)\,|)$$

$$\leqq \epsilon I_y J_x h(y)f'(x) = \epsilon I(h)J(f').$$

Similarly, if $g \in L'$ and h is symmetric and suitably restricted, then

$$|\, I(h)J(g) - J(h)I(g)\,| \leqq \epsilon I(h)J(g'),$$

where g' is fixed, and has the same relation to g that f' has to f. Thus

$$\left|\, \frac{J(f)}{I(f)} - \frac{J(g)}{I(g)} \,\right| \leqq \epsilon \left|\, \frac{J(f')}{I(f)} + \frac{J(g')}{I(g)} \,\right|$$

and, since ϵ is arbitrary, the left member is zero and the ratios are equal. Thus if g_0 is fixed and $c = J(g_0)/I(g_0)$, then $J(f) = cI(f)$ for all $f \in L^+$, completing the proof of the theorem.

If I is known to be right invariant as well as left invariant, as is the case, for instance, if G is commutative, then the uniqueness proof is trivial. For if f and $h \in L^+$ and $h^*(x) = h(x^{-1})$, then

$$J(h)I(f) = J_y I_x h(y)f(x) = J_y I_x h(y)f(xy) = I_x J_y h(y)f(xy)$$

$$= I_x J_y h(x^{-1}y)f(y) = J_y I_x h^*(y^{-1}x)f(y)$$

$$= J_y I_x h^*(x)f(y) = I(h^*)J(f),$$

so that $J(f) = cI(f)$ where $c = J(h)/I(h^*)$.

29E. Theorem. *G is compact if and only if $\mu(G) < \infty$ (the constant functions are summable).*

Proof. If G is compact, then $1 \in L$, and $\mu(G) = I(1) < \infty$. Conversely, if G is not compact and if V is a neighborhood of the identity with compact closure, then G cannot be covered by a finite set of translates of V. Therefore we can choose a sequence

$\{p_n\}$ of points in G such that $p_n \notin \bigcup_1^{n-1} p_i V$. Now let U be a symmetric neighborhood such that $U^2 \subset V$. Then the open sets $p_n U$ are all disjoint, for if $m > n$ and $p_n U \cap p_m U \neq \varnothing$, then $p_m \in p_n U^2 \subset p_n V$, a contradiction. Since the common measure of the open sets $p_n U$ is positive, it follows that the measure of G is infinite, and the constant functions are not summable, q.e.d.

If G is compact, Haar measure is customarily normalized so that $\mu(G) = 1 \left(\int 1 \, d\mu = I(1) = 1 \right)$.

§ 30. THE MODULAR FUNCTION

30A. Left invariant Haar measure need not be also right invariant—in general $I(f^t) \neq I(f)$. However, if t is fixed, $I(f^t)$ is a left invariant integral, $I(f_s{}^t) = I(f^t{}_s) = I(f^t)$, and because of the uniqueness of Haar measure there exists a positive constant $\Delta(t)$ such that

$$I(f^t) \equiv \Delta(t)I(f).$$

The function $\Delta(t)$ is called the *modular function* of G; if $\Delta(t) \equiv 1$, so that Haar measure is both left and right invariant, G is said to be *unimodular*.

Lemma. *If G is Abelian or compact, then G is unimodular.*

Proof. The Abelian case is trivial. If G is compact and if $f \equiv 1$, then $\Delta(s) = \Delta(s)I(f) = I(f^s) = I(1) = 1$, so that the compact case is also practically trivial.

We have already observed earlier that, if $f \in L$, then $I(f^t)$ is a continuous function of t. Thus $\Delta(t)$ is continuous. Also $\Delta(st)I(f) = I(f^{st}) = I((f^s)^t) = \Delta(t)I(f^s) = \Delta(s)\Delta(t)I(f)$. Thus $\Delta(t)$ *is a continuous homomorphism of G into the positive real numbers*.

30B. As might be expected, Haar measure is not ordinarily inverse invariant, and the modular function again plays its adjusting role. We use the customary notation $\int f(x) \, dx$ instead of $I(f)$ below because of its greater flexibility in exhibiting the variable.

Theorem. $\int f(x^{-1}) \, \Delta(x^{-1}) \, dx = \int f(x) \, dx.$

Before starting the proof we define the function f^* by

$$f^*(x) = \overline{f(x^{-1})} \, \Delta(x^{-1}).$$

The complex conjugate is taken so that, at the appropriate moment, the mapping $f \to f^*$ will be seen to be an involution. The formulas

$$f^{*s} = \Delta(s)f_s{}^*, \quad f_s{}^* = \Delta(s)f^*{}_s$$

follow directly from the definition.

Proof. In proof of the theorem we first observe that the integral above, $J(f) = I(f^{*-})$, is left invariant: $J(f_s) = I(f_s{}^{*-}) = \Delta(s^{-1})I(f^{*-s}) = I(f^{*-}) = J(f)$. Therefore $J(f) = cI(f)$. But now, given ϵ, we can choose a symmetric neighborhood V of e on which $| 1 - \Delta(s) | < \epsilon$ and then choose a symmetric $f \in L_V{}^+$ such that $I(f) = 1$, giving

$$| 1 - c | = | (1 - c)I(f) | = | I(f) - J(f) |$$
$$= | I((1 - \Delta^{-1})f) | < \epsilon I(f) = \epsilon.$$

Since ϵ is arbitrary, $c = 1$, q.e.d.

Corollary 1. $f^* \in L^1$ *if and only if* $f \in L^1$, *and* $\| f^* \|_1 = \| f \|_1$. This follows from the theorem and the fact that L is dense in L^1.

Corollary 2. *Haar measure is inverse invariant if and only if* G *is unimodular.*

30C. The following theorem does not really belong here, but there seems to be no better spot for it.

Theorem. *If* $g \in L^p$ $(1 \leqq p < \infty)$, *then* g_s *and* g^s, *as elements of* L^p, *are continuous functions of* s.

Proof. Given ϵ, we choose $f \in L$ so that $\| g - f \|_p = \| g_s - f_s \|_p < \epsilon/3$ and by **28B** we choose V so that $\| f - f_s \|_p < \epsilon/3$ if $s \in V$. Then

$$\| g - g_s \|_p \leqq \| g - f \|_p + \| f - f_s \|_p + \| f_s - g_s \|_p < \epsilon$$

if $s \in V$, showing that g_s is a continuous function of s at $s = e$.

Since $|| f^s - g^s ||_p = \Delta(s)^{1/p} || f - g ||_p$, the above inequality remains valid when g_s and f_s are replaced by g^s and f^s if the choice of V is now modified so that $|| f - f^s ||_p < \epsilon/6$ and $\Delta(s)^{1/p} < 2$ when $s \in V$. The equations $|| f_{sx} - f_x ||_p = || f_s - f ||_p$ and $|| f^{sx} - f^x ||_p = \Delta(x)^{1/p} || f^s - f ||_p$ show that both functions are continuous everywhere, and, in fact, that f_s is left uniformly continuous.

30D. The classes of groups treated in detail in this book, Abelian and compact groups, are unimodular. However, many important groups are not unimodular, and it will be worth while to exhibit a class of such groups. Let G and H be locally compact unimodular groups and suppose that each $\sigma \in G$ defines an automorphism of H (that is, G is mapped homomorphically into the automorphism group of H), the result of applying σ to an element $x \in H$ being designated $\sigma(x)$. The reader can check that the Cartesian product $G \times H$ becomes a group if multiplication is defined by

$$\langle \sigma_1, x_1 \rangle \langle \sigma_2, x_2 \rangle = \langle \sigma_1 \sigma_2, \sigma_2(x_1)x_2 \rangle.$$

Furthermore, if $\sigma(x)$ is continuous in the two variables σ and x simultaneously, then this group is locally compact in the ordinary Cartesian product topology. Such a group is called a semi-direct product of H by G. It contains H as a normal subgroup and each coset of the quotient group contains exactly one element of G.

If A and B are compact subsets of G and H respectively, then $\langle \sigma_0, x_0 \rangle (A \times B)$ is the set of all pairs $\langle \sigma_0 \sigma, \sigma(x_0)x \rangle$ such that $\sigma \in A$ and $x \in B$. If we compute the Cartesian product measure of this set by the Fubini theorem, integrating first with respect to x, we get $\mu(A)\nu(B)$, where μ and ν are the Haar measures in G and H respectively. This is also the product measure of $A \times B$. Thus the Cartesian product measure is the left invariant Haar measure for the semi-direct product group.

To compute the modular function Δ we consider $(A \times B)\langle \sigma_0, x_0 \rangle = (A\sigma_0) \times (\sigma_0(B)x_0)$, whose product measure is $\mu(A)\nu(\sigma_0(B))$. If the automorphism σ_0 multiplies the Haar measure in H by a factor $\delta(\sigma_0)$, then the above measure is $\delta(\sigma_0)\mu(A)\nu(B)$. The modular function $\Delta(\langle \sigma, x \rangle)$ therefore has the value $\delta(\sigma)$, and the semi-direct product is unimodular only if $\delta \equiv 1$.

As a simple example we take $H = R$, the additive group of the real numbers, and G as the multiplicative group of the positive reals, with $\sigma(x) = \sigma \cdot x$. Thus $\langle a, x \rangle \langle b, y \rangle = \langle ab, bx + y \rangle$ and $\Delta(\langle a, x \rangle) = a$. This group is, therefore, not unimodular.

30E. It is clear that, if G and H are not unimodular, but have modular functions Δ_1 and Δ_2 respectively, then the above calculation will give the modular function Δ in terms of Δ_1 and Δ_2. Cartesian product measure is seen to be the left invariant Haar measure exactly as above, but now the product measure of $(A \times B)\langle \sigma_0, x_0 \rangle = (A\sigma_0) \times (\sigma_0(B)x_0)$ is $\delta(\sigma_0)\Delta_1(\sigma_0)\Delta_2(x_0)$ $[\mu(A)\nu(B)]$ so that

$$\Delta(\langle \sigma, x \rangle) = \delta(\sigma)\Delta_1(\sigma)\Delta_2(x).$$

If in this situation we take H as the additive group of the real numbers (so that $\Delta_2 \equiv 1$) and $\sigma(x) = x[\Delta_1(\sigma)]^{-1}$, then $\delta(\sigma) = \Delta_1(\sigma)^{-1}$ and $\Delta(\langle \sigma, x \rangle) \equiv 1$. Thus any locally compact group G can be enlarged slightly to a unimodular group which is a semi-direct product of the reals by G. This was pointed out by Gleason [18].

§ 31. THE GROUP ALGEBRA

In this section we prove the existence of the convolution operation and show that with it as multiplication $L^1(G)$ becomes a Banach algebra. **31C–G** then establish some of the special properties of this algebra which follow more or less immediately from its definition.

31A. We define the *convolution* of f with g, denoted $f * g$, by

$$[f * g](x) = \int f(xy)g(y^{-1}) \, dy = \int f(y)g(y^{-1}x) \, dy,$$

the equality of the integrals being assured by the left invariance of the Haar integral. The ordinary convolution integral on the real line, $\int_{-\infty}^{\infty} f(x - y)g(y) \, dy$, is obviously a special case of this definition. It also agrees with and generalizes the classical notion used in the theory of finite groups. A finite group G is compact if it is given the discrete topology, and the points of G, being congruent non-void open sets, must have equal positive Haar measures. If the Haar measure of G is normalized by giving each point the measure 1, and if f and g are any two complex-valued functions on G, then

$$f * g(x) = \int f(xy)g(y^{-1})\, dy = \sum_y f(xy)g(y^{-1}) = \sum_{uv=x} f(u)g(v)$$

which is the classical formula for convolution.

The above definition is purely formal and we must start by showing that $f * g$ exists.

Lemma. *If $f \in L_A$ and $g \in L_B$, then $f * g \in L_{AB}$.*

Proof. The integrand $f(y)g(y^{-1}x)$ is continuous as a function of y for every x and is zero unless $y \in A$ and $y^{-1}x \in B$. Thus $f * g(x)$ is defined for every x and is zero unless $x \in AB$. Also $| f * g(x_1) - f * g(x_2) | \leq \| f_{x_1} - f_{x_2} \|_\infty \cdot \int | g(y^{-1}) |\, dy$ and since f_x is by **28B** continuous in the uniform norm as a function of x it follows that $f * g$ is continuous.

Theorem. *If f, $g \in \mathcal{B}^+$, then $f(y)g(y^{-1}x) \in \mathcal{B}^+(G \times G)$ and $\| f * g \|_p \leq \| f \|_1 \cdot \| g \|_p$, where $1 \leq p \leq \infty$.*

Proof. Let $f \circ g$ be the function $f(y)g(y^{-1}x)$. If $f \in L^+$, then the family of functions $g \in \mathcal{B}^+$ such that $f \circ g \in \mathcal{B}^+(G \times G)$ is L-monotone and includes L^+, and is therefore equal to $\mathcal{B}^+$. Thus if g is an L-bounded function of $\mathcal{B}^+$, then the family of functions $f \in \mathcal{B}^+$ such that $f \circ g \in \mathcal{B}^+$ includes L^+ and is L-monotone, and therefore equals $\mathcal{B}^+$. In particular $f \circ g \in \mathcal{B}^+$ whenever f and g are L-bounded functions of $\mathcal{B}^+$. The general assertion then follows from the fact that any non-negative Baire function is the limit of an increasing sequence of L-bounded Baire functions. The Fubini theorem implies that $f(y)g(y^{-1}x)$ is integrable in y for every x, that $(f * g)(x) = \int f(y)g(y^{-1}x)\, dy$ is integrable, and that $\| f * g \|_1 = \iint f(y)g(y^{-1}x)\, dx\, dy = \| f \|_1 \cdot \| g \|_1$. If $h \in \mathcal{B}^+$ then also $f(y)g(y^{-1}x)h(x) \in \mathcal{B}^+(G \times G)$ (see **13C**) and, as above,

$$(f * g, h) = \int f(y)\left[\int g(y^{-1}x)h(x)\, dx \right] dy \leq \| f \|_1 \| g \|_p \| h \|_q$$

by the general Hölder inequality **14C**.

It follows, using **15C**, that $\| f * g \|_p \leqq \| f \|_1 \| g \|_p$, and, in particular, that $f * g \in L^p$ if $f \in L^1$ and $g \in L^p$. The case $p = \infty$ is obvious.

Corollary. *If* $f \in L^1$ *and* $g \in L^p$, *then* $f(y)g(y^{-1}x)$ *is summable in* y *for almost all* x, $(f * g)(x) \in L^p$ *and* $\| f * g \|_p \leqq \| f \|_1 \| g \|_p$.

Proof. The sixteen convolutions obtained by separating each of f and g into its four non-negative parts belong to L^p by the above theorem, and their sum, which is equal to $\int f(y)g(y^{-1}x)\,dx$ wherever all sixteen are finite, is thus an element of L^p (with the usual ambiguity about its definition at points where the summands assume opposite infinities as values). Moreover, $\| f * g \|_p \leqq \| \, | f | * | g | \, \|_p \leqq \| f \|_1 \| g \|_p$, by the above theorem.

31B. Theorem. *Under convolution as multiplication* $L^1(G)$ *forms a Banach algebra having a natural continuous involution* $f \to f^*$.

Proof. We have seen earlier that the mapping $f \to f^*$ of L^1 onto itself is norm preserving, and we now complete the proof that it is an involution. It is clearly additive and conjugate linear. Also

$$(f * g)^*(x) = \int \overline{f(x^{-1}y)}\ \overline{g(y^{-1})}\Delta(x^{-1})\,dy$$

$$= \int \overline{g(y^{-1})}\Delta(y^{-1})\overline{f((y^{-1}x)^{-1})}\Delta((y^{-1}x)^{-1})\,dy$$

$$= (g^* * f^*)(x),$$

so that $(f * g)^* = g^* * f^*$.

The associativity of convolution depends on the left invariance of the Haar integral. If $f, g, h \in \mathfrak{B}^+$, we have

$$((f * g) * h)(x) = \int [(f * g)(xy)]h(y^{-1})\,dy$$

$$= \iint f(xyz)g(z^{-1})h(y^{-1})\,dy\,dz$$

$$= \iint f(xz)g(z^{-1}y)h(y^{-1})\,dy\,dz$$

$$= \int f(xz)[(g * h)(z^{-1})]\,dz = (f * (g * h))(x).$$

The associative law is therefore valid for any combination of functions from the L^p classes for which the convolutions involved are all defined. The convolution $f * g$ is also clearly linear in f and g. Since L^1 is closed under convolution and $\| f * g \|_1 \leq \| f \|_1 \cdot \| g \|_1$ by **31A**, it follows that L^1 is a Banach algebra with convolution as multiplication. It is called the group algebra (or group ring) of G, and generalizes the notion of group algebra used in the classical theory of finite groups.

We prove below a number of useful elementary properties of the group algebra $L^1(G)$.

31C. Theorem. $L^1(G)$ *is commutative if and only if G is commutative.*

Proof. If G is commutative, then G is also unimodular and

$$f * g = \int f(y)g(y^{-1}x)\, dy = \int g(xy)f(y^{-1})\, dy = g * f.$$

Now suppose that $L^1(G)$ is commutative and that f, $g \in L$. Then

$$0 = f * g - g * f = \int [f(xy)g(y^{-1}) - g(y)f(y^{-1}x)]\, dy$$

$$= \int g(y)[f(xy^{-1})\Delta(y^{-1}) - f(y^{-1}x)]\, dy.$$

Since this holds for every g in L, it follows that

$$f(xu)\, \Delta(u) - f(ux) \equiv 0.$$

Taking $x = e$ we have $\Delta(u) \equiv 1$, so that $f(xu) - f(ux) = 0$ for every f in L. Since L separates points, it follows that $xu - ux = 0$ and G is commutative.

31D. Theorem. $L^1(G)$ *has an identity if and only if G is discrete.*

Proof. If G is discrete, the points of G are congruent open sets having equal positive Haar measures which may be taken as 1. Then $f(x)$ is summable if and only if $f(x) = 0$ except on a countable set $\{x_n\}$ and $\sum | f(x_n) | < \infty$. The function $e(x)$ which is 1 at $x = e$ and zero elsewhere is an identity:

$$f * e(x) = \int f(y)e(y^{-1}x)\, dy = \sum_y f(y)e(y^{-1}x) = f(x).$$

Conversely suppose that $u(x)$ is an identity for L. We show that there is a positive lower bound to the measures of non-void open (Baire) sets. Otherwise, given any ϵ there exists an open neighborhood V of the identity e whose measure is less than ϵ, and hence one such that $\int_V |\, u(x)\,|\, dx < \epsilon$. Choose a symmetric U so that $U^2 \subset V$ and let f be its characteristic function. Then

$$f(x) = (u * f)(x) = \int u(y)f(y^{-1}x)\, dy = \int_{xU} u(y)\, dy \leqq \int_V |\, u\,| < \epsilon$$

for almost all x in U, contradicting $f(x) \equiv 1$ in U. Therefore, there is a number $a > 0$ such that the measure of every non-void open Baire set is at least a. From this it follows at once that every open set whose closure is compact, and which therefore has finite measure, contains only a finite set of points, since otherwise its measure is seen to be $\geqq na$ for every n by choosing n disjoint non-void open subsets. Therefore, every point is an open set, and the topology is discrete.

31E. In any case $L^1(G)$ has an "approximate identity."

Theorem. *Given $f \in L^p$ ($1 \leqq p < \infty$) and $\epsilon > 0$, there exists a neighborhood V of the identity e such that $\|\, f * u - f\,\|_p < \epsilon$ and $\|\, u * f - f\,\|_p < \epsilon$ whenever u is any function of L^{1+} such that $u = 0$ outside of V and $\int u = 1$.*

Proof. If $h \in L^q$ the Fubini theorem and Hölder inequality imply that

$$|\, (u * f - f, h)\,| = \left|\, \iint u(y)(f(y^{-1}x) - f(x))\overline{h(x)}\, dy\, dx\,\right|$$

$$\leqq \|\, h\,\|_q \int \|\, f_{y^{-1}} - f\,\|_p u(y)\, dy.$$

Thus $\|\, u * f - f\,\|_p \leqq \int \|\, f_{y^{-1}} - f\,\|_p u(y)\, dy$. If V is chosen (by **30C**) so that $\|\, f_{y^{-1}} - f\,\|_p < \epsilon$ whenever $y \in V$ and if $u \in (L^1)_V{}^+$, then $\|\, u * f - f\,\|_p < \epsilon \int u = \epsilon$.

The proof for $f * u$ is similar but somewhat complicated by modularity. If $m = \int u(x^{-1})\, dx$, then

$$| (f * u - f, h) | = | \iint (f(xy) - f(x)/m)u(y^{-1})\overline{h(x)} \, dy \, dx |$$

$$\leq \| h \|_q \int \| mf^{y^{-1}} - f \|_p (u(y^{-1})/m) \, dy$$

and $\| f * u - f \|_p \leq \int \| mf^{y^{-1}} - f \|_p u(y^{-1})/m \, dy$. Now $m \to 1$ as V decreases. (For $m = \int [u(x^{-1}) \, \Delta(x^{-1})] \, \Delta(x) \, dx$, where $\int u(x^{-1}) \, \Delta(x^{-1}) \, dx = \int u(x) = 1$, $\Delta(x)$ is continuous and $\Delta(e) = 1$.) Thus there exists V such that $\| mf^{y^{-1}} - f \|_p < \epsilon$ if $y^{-1} \in V$ and $u \in (L^1)_V{}^+$, giving $\| f * u - f \|_p < \epsilon \int [u(y^{-1})/m] \, dy = \epsilon$.

The neighborhoods of the identity V form a directed system under inclusion, and if u_V is a non-negative function vanishing off V and satisfying $\int u_V = 1$, then $u_V * f$ and $f * u_V$ converge to f in the pth norm for any $f \in L^p$ by the above theorem. The directed system of functions u_V can be used in place of an identity for many arguments and is called an *approximate identity*.

31F. Theorem. *A closed subset of L^1 is a left (right) ideal if and only if it is a left (right) invariant subspace.*

Proof. Let I be a closed left ideal and let u run through an approximate identity. If $f \in I$, then $u_x * f \in I$. But $u_x * f = (u * f)_x \to f_x$, since $u * f \to f$, and therefore $f_x \in I$. Thus every closed left ideal is a left invariant subspace.

Now let I be a closed left invariant subspace of L^1. If $I^\perp$ is the set of all $g \in L^\infty$ such that $(f, g) = 0$ for every $f \in I$, then we know (see **8C**) that $I = (I^\perp)^\perp$, i.e., that, if $f \in L^1$, then $f \in I$ if and only if $(f, g) = 0$ for every $g \in I^\perp$. But if $h \in L^1, f \in I$ and $g \in I^\perp$, then $(h * f, g) = \iint h(y)f(y^{-1}x)\overline{g(x)} \, dy \, dx = \int h(y)$ $\int [f(y^{-1}x)\overline{g(x)} \, dx] \, dy = 0$ (since $f_y \perp g$ for every y), which proves by the above remark that $h * f \in I$. That is, I is a left ideal.

The same proof, with only trivial modifications, works for right ideals.

31G. We conclude this section with some remarks about positive definite functions. For this discussion we restrict ourselves exclusively to unimodular groups. It will be remembered that the notion of positivity was introduced in connection with an abstract Banach algebra having a continuous involution. The algebra which we have before us now is $L^1(G)$, and we start with the ideal L^0 consisting of the uniformly continuous functions of L^1, and the function φ on L^0 defined by $\varphi(f) = f(e)$. Then

$$\varphi(f * f^*) = \int f(x)\overline{f(x)}\, dx = \| f \|_2^2 \geqq 0.$$ Thus φ is positive and H_φ is $L^2(G)$. Since $\| g * f \|_2 \leqq \| g \|_1 \cdot \| f \|_2$, the operators U_g defined by $U_g f = g * f$ are bounded and the mapping $g \rightarrow U_g$ is a $*$-representation of $L^1(G)$. It is called the left regular representation and will be discussed further in **32D.** The functional φ is thus unitary, but is clearly not continuous (in the L^1 norm).

At the other extreme are the positive functionals which are continuous in the L^1 norm.

Lemma. *A positive functional on $L^1(G)$ is continuous if and only if it is extendable.*

Proof. We have seen earlier (**26I**) that any extendable positive functional is automatically continuous. The converse implication follows here because of the existence of an approximate identity. Given a continuous positive functional P, we have

$$| P(f) |^2 = \lim | P(f * u) |^2 \leqq P(f * f^*) \overline{\lim} P(u^* * u)$$

$$\leqq \| P \| P(f * f^*),$$

$$P(f^*) = \lim P(f^* * u) = \lim \overline{P(u^* * f)} = \overline{P(f)},$$

proving the lemma.

Now every continuous functional P on L^1 is given by a function $p \in L^\infty$, and, if P is positive, p is called *positive definite*. Notice that then $p = p^*$, for $(f, p) = P(f) = \overline{P(f^*)} = \overline{(f^*, p)} = (p, f^*) = (f, p^*)$ for every $f \in L^1$. This notion of positive definiteness is formally the same as that introduced in **26J**, the fixed unitary functional φ being that considered in the first paragraph above: $P(f) = (f, p) = (f, p^*) = (p, f^*) = [p * f](e) = \varphi(p * f)$.

The condition that a function $p \in L^{\infty}$ be positive definite can be written

$$\iint f(x)\overline{f(y)}p(xy^{-1})\,dx\,dy \geqq 0$$

for every $f \in L^1$.

§ 32. REPRESENTATIONS

32A. A *representation* T of G is a strongly continuous homomorphism of G onto a group of linear transformations on a complex vector space X. That is, if T_s is the transformation associated with the group element s, then $T_{st} = T_s T_t$ and $T_s(x)$ is a continuous function of s for every $x \in X$. If X is finite dimensional, this requirement of continuity can be expressed as the continuity of the coefficients in the matrix for T_s when a fixed basis has been chosen for X. When X is infinite dimensional some kind of topology for X must be specified. Usually X is a Banach space and here we shall confine ourselves to reflexive Banach spaces and Hilbert space. T will be said to be *bounded* if there is a uniform bound to the norms $\| T_s \|$, $s \in G$. T is a *unitary* representation if X is a Hilbert space and the transformations T_s are all unitary.

32B. *If T is a bounded representation of G on a reflexive Banach space X and if T_f is defined as $\int f(x) T_x\, dx$ for every $f \in L^1$, then the mapping $f \to T_f$ is a bounded representation of $L^1(G)$. If T is unitary, then the integrated representation is a $*$-representation of $L^1(G)$.*

Proof. The function

$$F(f, x, y) = \int f(s)(T_s x, y)\,ds = \int f(s) y(T_s x)\,ds$$

is trilinear and satisfies $| F(f, x, y) | \leqq \| f \|_1 B \| x \| \| y \|$. It follows exactly as in **26F** that there is a uniquely determined linear transformation T_f such that $(T_f x, y) = \int f(s)(T_s x, y)\,ds$, $\| T_f \| \leqq B \| f \|_1$, and the mapping $f \to T_f$ is linear. Moreover,

$$(T_{f*g}x, y) = \iint f(t)g(t^{-1}s)(T_s x, y)\, dt\, ds$$

$$= \iint f(t)g(s)(T_{ts}x, y)\, ds\, dt$$

$$= \int g(s)(T_f T_s x, y)\, ds\, dt = \int g(s)(T_s x, (T_f)^* y)\, ds$$

$$= \int (T_g x, (T_f)^* y) = (T_f T_g x, y).$$

Thus $T_{f*g} = T_f T_g$ and the mapping $f \to T_f$ is a homomorphism. If X is a Hilbert space and T is a unitary representation, then

$$(T_{f*}x, y) = \int \overline{f(s^{-1})}\Delta(s^{-1})(T_s x, y)\, ds = \left(\int f(s)(T_s y, x)\, ds\right)^{-}$$

$$= \overline{(T_f y, x)} = (x, T_f y).$$

Thus $(T_f)^* = T_{f*}$ and T is a $*$-representation.

Remark: No non-zero element of X is annihilated by every T_f. For, given $0 \neq x \in X$ and $y \in X^*$ such that $(x, y) \neq 0$ we can find a neighborhood V of the identity in G such that $(T_s x, y)$ differs only slightly from (x, y) if $s \in V$. If f is the characteristic function of V, then $(T_f x, y) = \int_V (T_s x, y)\, ds$ differs only slightly from $\mu(V)(x, y)$, and hence $T_f x$ is not zero.

32C. Theorem. *Conversely, a bounded representation T of $L^1(G)$ over a Banach space X arises in the above way if the union of the ranges of the operators $T_f, f \in L^1$, is dense in X. If X is a Hilbert space and T is a $*$-representation, then T is unitary.*

Proof. Let X_r be the union of the ranges of all the operators $T_f(f \in L^1)$; X_r is dense in X by hypothesis. Let u run through an approximate identity of L^1. We know that $u_a * f = (u * f)_a \to f_a$, and therefore $\| T_{u_a}T_f - T_{f_a} \| \to 0$. Thus T_{u_a} converges strongly (that is, pointwise) on X_r to an operator U_a satisfying $U_a T_f = T_{f_a}$, and since $\| T_{u_a} \| \leq B\| u_a \|_1 = B$ for every u, it follows that $\| U_a \| \leq B$ and that U_a is uniquely defined on the whole of X (see **7F**). Since $U_{ab}T_f = T_{f_{ab}} = T_{(f_a)_b} = U_b T_{f_a} = U_b U_a T_f$ (i.e., $U_{ab} = U_b U_a$ on X_r), we have $U_{ab} = U_b U_a$ and the mapping $a \rightarrow U_a$ is therefore an anti-homomorphism. Finally, f_a is a continuous function of a as an element of L_1, and therefore $U_a T_f = T_{f_a}$ is a continuous function of a. Thus U_a is strongly continuous on X_r, and hence on X. U is an anti-homomorphism, and we set $T_a = U_{a^{-1}}$ to get a direct homomorphism.

It remains to be shown that $(T_f x, y) = \int f(s)(T_s x, y) \, ds$ for every $x \in X$, $y \in X^*$ and $f \in L^1$. We may clearly restrict x to any dense subset of X and f to L. Thus, it is sufficient to show that $(T_{f \ast g} z, y) = \int f(s)(T_s T_g z, y) \, ds$ for all $f, g \in L$. Now if x and y are fixed, the linear functional $J(f) = (T_f x, y)$ satisfies $|J(f)| \leqq B\| x \| \| y \| \| f \|_1$, and is therefore a complex-valued integral (the sum of its four variations). We therefore have from the Fubini theorem that

$$(T_{f \ast g} x, y) = J_t \left[\int f(s) g(s^{-1}t) \, ds \right] = \int f(s)[J_t g(s^{-1}t)] \, ds$$

$$= \int f(s)(T_s T_g x, y) \, ds$$

as desired.

If X is a Hilbert space and T is a $*$-representation, we want to show that T_s is unitary for every s, i.e., that $(T_s)^* = (T_s)^{-1} = T_{s^{-1}}$. But $(T_s)^* = T_{s^{-1}} \Leftrightarrow (T_s)^* T_f = T_{s^{-1}} T_f = T_{f_s}$ for all $f \in L^1$ $\Leftrightarrow (u_{s^{-1}})^* \ast f \rightarrow f_s$ as u runs through an approximate identity $\Leftrightarrow f^* \ast u_{s^{-1}} \rightarrow f_s^* \Leftrightarrow g \ast u_{s^{-1}} \rightarrow g^*_s{}^*$. Now by direct computation $g^*{}_s{}^* = \Delta(s^{-1}) g^s$ and $g \ast u_{s^{-1}} = \Delta(s^{-1}) g^s \ast u$. Since we know that $g^s \ast u \rightarrow g^s$, the result follows.

32D. If we take X as the Banach space $L^p(G)$ for some fixed p in $[1, \infty]$ and define the operator T_f on X by $T_f g = f \ast g$ for every $f \in L^1$, $g \in L^p$, then we can check directly that the mapping $f \rightarrow T_f$ is a bounded representation of the algebra $L^1(G)$. For instance, $T_f T_g h = f \ast (g \ast h) = (f \ast g) \ast h = T_{f \ast g} h$, giving $T_{f \ast g} = T_f T_g$. Moreover the inequality $\| f \ast g \|_p \leqq \| f \|_1 \| g \|_p$ implies that $\| T_f \| \leqq \| f \|_1$, so that the representation has the bound 1.

There is also an obvious group representation $s \rightarrow T_s$ defined by $T_s f = f_{s^{-1}}$, or $[T_s f](t) = f(s^{-1}t)$. The transformations T_s are all isometries by the left invariance of Haar measure.

This pair of representations is interconnected in exactly the same way as in the above two theorems. Thus if $g \in L^p$ and $h \in L^q$ $(1/p + 1/q = 1)$, then

$$\int f(s)(T_s g, h) \, ds = \iint f(s)g(s^{-1}t)\overline{h(t)} \, ds \, dt = (f * g, h).$$

If $p = 2$, the transformations T_s are unitary by the left invariance of Haar measure: $(T_s f, T_s g) = \int f(s^{-1}t)g(s^{-1}t) \, dt = \int f(t)g(t) \, dt = (f, g)$. It follows from our general theory that the representation of L^1 is a *-representation. Of course, this fact can also be checked directly—the equality $(f * g, h) = (g, f^* * h)$ can be computed directly from the Fubini theorem and the left invariance of the Haar integral. These representations are called the (left) *regular* representations of G and $L^1(G)$ respectively. They are both faithful representations (one-to-one mappings), for if $s \neq e$ we can choose g to vanish outside a small enough neighborhood of e so that $gg_s = 0$, giving in particular $T_s g \neq g$ in $L^2(G)$, and if $f \neq 0$ then we can choose $u \in L^1 \cap L^2$ similarly so that $\|f * u - f\|_2 < \epsilon \|f\|_2$, giving in particular that $T_f u = f * u \neq 0$ in $L^2(G)$.

§ 33. QUOTIENT MEASURES

In this section we consider invariant measure on quotient groups, and, more generally, quotient spaces, and prove the Fubini-like theorem relating such quotient measures to the invariant measures on the whole group and kernel subgroup. This material is not needed for most of the subsequent discussion, and since it gets fairly technical the reader may wish to omit it. The method of procedure is taken quite directly from Weil [48].

33A. Let H be a fixed closed subgroup of G, and let I and J be respectively the left invariant Haar integrals of G and H. It is understood that $J(f)$ is the integral over H of the restriction of f to H. If $f \in L(G)$, it follows from the uniform continuity of f that the function f' defined by $f'(x) = J(f_x) = J_t f(xt)$ is continuous. Also $f'(xs) = f'(x)$ for every $s \in H$ by the left invariance of J on H, so that f' is constant on the left cosets of H.

If $f \in L_C$, then the related function F on G/H defined by $F(\alpha(x)) = f'(x) = J(f_x) = J_t f(xt)$ belongs to $L_{\alpha(C)}$. The value $F(\alpha(x)) = f'(x)$ can also be obtained as the integral of f over the coset containing x with respect to the Haar measure of H transplanted to this coset.

If H is a closed *normal* subgroup, then G/H is a group and has a left invariant Haar integral K. $K_x J_t f(xt)$ is understood to be $K(F)$ where $F(\alpha(x)) = J_t f(xt)$. This is equivalent to transferring the domain of K to those continuous functions on G which are constant on the cosets of H and each of which vanishes off some set of the form CH, where C is compact. If g is such a function, then so is g_y and the left invariance of K becomes simply $K(g) = K(g_y)$ for all $y \in G$. Then $K_x J_t f(xt)$ is a left invariant integral on $L(G)$, $K_x J_t f_y(xt) = K_x f_y'(x) = K_x f'(x) = K_x J_t f(xt)$, and $K_x J_t f(xt) = kI(f)$. We are thus led to the following theorem.

Theorem. *If $f \in \mathcal{B}^+(G)$, then $f_x \in \mathcal{B}^+(H)$ for every x and $J(f_x) = J_t f(xt) \in \mathcal{B}^+(K)$. If I, J and K are suitably normalized, then $K_x J_t f(xt) = I(f)$ for every $f \in \mathcal{B}^+(G)$.*

Proof. We have seen above that the family of non-negative functions for which the theorem is true includes L^+, with a suitable normalization for I. If the theorem holds for a sequence $\{f_n\}$ of L-bounded functions of $\mathcal{B}^+$ which converges monotonically to f, then $I(f) = \lim I(f_n) = \lim K_x J_t f_n(xt) = K_x(\lim J_t f_n(xt) = K_x J_t (\lim f_n(xt) = K_x J_t f(xt)$. Thus the family is L-monotone and hence equal to $\mathcal{B}^+(G)$, q.e.d.

Corollary. *If $f \in L^1(G)$, then $f_x \in L^1(H)$ for almost all x, $J_t f(xt) \in L^1(K)$ and $K_x J_t f(xt) = I(f)$.*

33B. We return now to the case where H is closed but not necessarily normal. The mapping $f \to F$ of $L(G)$ into $L(G/H)$ defined above is clearly linear, and we now show that it is onto.

Lemma. *If $F \in L^+(G/H)$, then there exists $f \in L^+(G)$ such that $F(\alpha(x)) = J(f_x)$.*

Proof. Let B be a compact subset of G/H such that $F \in L_B^+$, let A be a compact subset of G such that $\alpha(A) = B$ and choose $h \in L^+(G)$ such that $h > 0$ on A. Notice that $J(h_x) > 0$ if the coset containing x intersects A, i.e., if $x \in AH = \alpha^{-1}(B)$, and that $F(\alpha(x)) = 0$ in $(AH)' = \alpha^{-1}(B')$, an open set. Therefore, $g(x) = F(\alpha(x))/J(h_x)$ if $J(h_x) > 0$ and $g(x) = 0$ if $J(h_x) = 0$ defines an everywhere continuous function. Also $g(x)$ is constant on the cosets of H. Therefore $f = gh \in L^+(G)$ and $J_t f(xt) = g(x)J_t h(xt) = F(\alpha(x))$, q.e.d.

33C. If H is not normal, it is still true that every element $y \in G$ defines a homeomorphism of G/H onto itself: $\alpha(x) \to \alpha(yx)$, or $xH \to yxH$. If an integral K is invariant under all these homeomorphisms, then the theorem of **33A** goes through unchanged. More generally K may be *relatively invariant* in the sense that K is multiplied by a constant $D(y)$ under the homeomorphism induced by y: $K(F_y) = D(y)K(F)$, where, of course, $F_y(\alpha(x)) = F(\alpha(yx))$. The function $D(y)$ is called the *modulus*, or the *modular function*, of K, and, like Δ, is seen to be continuous and multiplicative.

Theorem. *If K is relatively invariant, with modulus D, then the functional $M(f) = K_x J_t D(xt) f(xt)$ is (essentially) the Haar integral $I(f)$, and all the conclusions of* **33A** *follow.*

Proof. $M(f_y) = K_x J_t D(xt) f(yxt) = D(y^{-1}) K_x [J_t D(xt) f(xt)]_y = K_x J_t D(xt) f(xt) = M(f)$. Thus $M(f)$ is left invariant on $L(G)$ and equals $I(f)$. The extensions to $\mathfrak{B}^+$ and to L^1 are the same as in **33A**.

33D. Any continuous homomorphism of G into the multiplicative group of the positive real numbers will be called a *real character*. Examples are the modular functions Δ and D.

Theorem. *In order that a real character D be the modular function for a relatively invariant measure K on the quotient space G/H, where H is a closed subgroup of G, with modular function δ, it is necessary and sufficient that $D(s) = \Delta(s)/\delta(s)$ for all $s \in H$.*

Proof. If D is the modular function of K, then $M(f) = I(f)$ as above and we have, for $s \in H$,

$$\Delta(s)M(f) = M(f^s) = K_x J_t D(xt) f(xts^{-1})$$
$$= D(s) K_x J_t D(xts^{-1}) f(xts^{-1})$$
$$= D(s)\, \delta(s) K_x J_t D(xt) f(xt) = D(s)\, \delta(s) M(f).$$

Thus $\Delta(s) = D(s)\, \delta(s)$ for every $s \in H$.

Now suppose, conversely, that $D(x)$ is a real character such that $\Delta(s) = D(s)\,\delta(s)$ for every $s \in H$, and define the functional K on $L(G/H)$ by $K_x(J(f_x)) = I(D^{-1}f)$. It follows from **33B** that this functional is defined on the whole of $L(G/H)$ once we have shown that the definition is unique, i.e., that, if $J(f_x) = 0$, then $I(D^{-1}f) = 0$. But if $J_t f(xt) = 0$ for all x and if $g \in L^+(G)$, then

$$0 = I_x J_t g(x) D^{-1}(x) f(xt) = J_t I_x\, \delta(t^{-1}) g(xt^{-1}) D^{-1}(x) f(x)$$

$$= I_x\{D^{-1}(x) f(x) J_t[\delta(t^{-1}) g(xt^{-1})]\} = I_x D^{-1}(x) f(x) J_t g(xt).$$

In order to prove that $I(D^{-1}f) = 0$ it is therefore sufficient to choose $g \in L^+(G)$ so that $J(g_x) = J_t g(xt) = 1$ for every x for which $f(x) \neq 0$. But if $f \in L_A$ and if a function is chosen from $L^+(G/H)$ which has the value 1 on the compact set $\alpha(A)$, then the required function g is constructed by the lemma of **33B**.

It remains to check that the functional K thus defined on $L^+(G/H)$ is relatively invariant with D as its modular function:

$$K_x J(f_{yx}) = I(D^{-1}f_y) = D(y) I((D^{-1}f)_y)$$

$$= D(y) I(D^{-1}f) = D(y) K_x J(f_x).$$

This completes the proof of the theorem.

33E. Returning to the case where H is normal we now see, since G/H has an invariant integral, with modular function $D \equiv 1$, that $\Delta(s) = \delta(s)$ for all $s \in H$. Taking H to be the normal subgroup on which $\Delta(t) = 1$, it follows that $\delta \equiv 1$ and H is unimodular. The real character Δ is itself a homomorphism having this group as kernel, and the quotient group is therefore isomorphic to a multiplicative subgroup of the positive real numbers. In a sense, therefore, every locally compact group is almost unimodular. Another sense in which this is true has been discussed in **30E**.

Chapter VII

LOCALLY COMPACT ABELIAN GROUPS

The L^1 group algebra of a locally compact Abelian group is a commutative Banach algebra with an involution, and much of the general theory of Chapter V is directly applicable. Thus we can take over the Bochner and Plancherel theorems whenever it seems desirable, and we can investigate questions of ideal theory such as the Wiener Tauberian theorem. We shall be especially concerned with reorienting this theory with respect to the character function (the kernel function of the Fourier transform), and much of our discussion will center around topological questions.

The character function is discussed and the character group defined in § 34 and various standard examples are given in § 35. § 36 is devoted to positivity, and includes the Bochner, Plancherel, and Stone theorems. A more miscellaneous set of theorems is gathered together in § 37, such as the Pontriagin duality theorem, the Wiener Tauberian theorem, and a simple form of the Poisson summation formula. The chapter concludes in § 38 with a brief discussion of compact Abelian groups; the general theory of compact groups is given in the next chapter.

§ 34.   THE CHARACTER GROUP

We know that each regular maximal ideal M in L^1 is the kernel of a homomorphism $f \to \hat{f}(M)$ of L^1 onto the complex numbers (see **23A, B**) and that this homomorphism, as an element of the conjugate space $(L^1)^*$ of L^1, is uniquely represented by a function $\alpha_M \in L^\infty$, $\hat{f}(M) = \int f(x)\overline{\alpha_M(x)}\, dx$ (see **15C, D**). We shall see below that α_M is (equivalent to) a continuous function, $|\alpha_M(x)| \equiv 1$ and $\alpha_M(xy) = \alpha_M(x)\alpha_M(y)$. Thus α_M is a continuous homomorphism of G into the multiplicative group of complex numbers of absolute value one; such a function is called a *character* of G. Moreover, every character arises in the above way. Thus the space $\mathfrak{M}$ of regular maximal ideals of L^1 is now identified with the set $\hat{G}$ of all the characters of G. Through this identification we know that $\hat{G}$, in the weak topology of $L^\infty = (L^1)^*$, is locally compact. And it follows rather easily that $\hat{G}$ is a locally compact group, multiplication being the pointwise multiplication of functions. The domain of the function $\hat{f}$ will be shifted from $\mathfrak{M}$ to $\hat{G}$ through the above identification, and, as a function on $\hat{G}$, $\hat{f}$ is now the classical *Fourier transform* of f.

We proceed to fill in the details.

34A. Theorem. *Given a fixed $M \in \mathfrak{M}$, let $f \in L^1$ be such that $\hat{f}(M) \neq 0$. Then the function α_M defined by $\alpha_M(x) = \hat{f}_x(M)/\hat{f}(M)$ is a character. It is independent of f, uniformly continuous, and continuous in the two variables x and M if the parameter M is also varied. If u runs through an approximate identity, then $\hat{u}_x(M)$ converges uniformly to $\alpha_M(x)$.*

Proof. That α_M is uniformly continuous follows from the inequality $|\hat{f}_x(M) - \hat{f}_y(M)| \leq \|f_x - f_y\|_1$ and the fact that f_x as an element of L^1 is a uniformly continuous function of x (**30C**). The inequality $|\hat{f}_x(M) - \hat{f}_{x_0}(M_0)| \leq \|f_x - f_{x_0}\|_1 + |\hat{f}_{x_0}(M) - \hat{f}_{x_0}(M_0)|$ shows that $\hat{f}_x(M)$ is continuous in the two variables x and M together, and the same holds for $\alpha_M(x)$ upon dividing by $\hat{f}(M)$.

The remaining properties of α_M are contained in the equation

$$\hat{f}(M)\hat{g}_x(M) = \hat{f}_x(M)\hat{g}(M)$$

(which comes from $f * g_x = f_x * g$). Thus $\hat{f}_x/\hat{f} = \hat{g}_x/\hat{g}$, proving that α is independent of f. Taking $g = f_y$ and dividing by $(\hat{f})^2$, we see that $\alpha_M(yx) = \alpha_M(y)\alpha_M(x)$. Thus α_M is a continuous homomorphism of G into the non-zero complex numbers. If $|\alpha_M(x_0)| > 1$, then $|\alpha_M(x_0^n)| = |\alpha_M(x_0)|^n \to \infty$. But α_M is bounded; therefore $\| \alpha_M \|_\infty \leqq 1$. Then $| \alpha_M{}^{-1}(x) | = | \alpha_M(x^{-1}) |$ $\leqq 1$, so that $| \alpha_M(x) | \equiv 1$. Thus α_M is a character.

Finally, taking $g = u$, we get $\hat{u}_x(M) = k\alpha_M(x)$, where $k = \hat{u}(M)$, and since $\hat{u}(M) \to 1$ as u runs through an approximate identity $(\| \hat{u}f - \hat{f} \|_\infty \leqq \| u * f - f \|_1 \to 0)$ it follows that $\hat{u}_x(M)$ converges uniformly to $\alpha_M(x)$.

34B. Theorem. *The mapping $M \to \alpha_M$ is a one-to-one mapping of $\mathfrak{M}$ onto the set of all characters of G, and*

$$\hat{f}(M) = \int f(x)\overline{\alpha_M(x)} \, dx.$$

Proof. We first derive the formula. Given M, the homomorphism $f \to \hat{f}(M)$ is in particular a linear functional on L^1, and therefore there exists $\alpha \in L^\infty$ such that $\hat{f}(M) = \int f(x)\overline{\alpha(x)} \, dx$. Then

$$\int f(x)\overline{\alpha_M(x)} \, dx = \lim_u \int f(x)u_{x^{-1}}(M) \, dx$$

$$= \lim_u \iint f(x)u(x^{-1}y)\overline{\alpha(y)} \, dx \, dy$$

$$= \lim_u \int (f * u)(y)\overline{\alpha(y)} \, dy$$

$$= \int f(y)\overline{\alpha(y)} \, dy = \hat{f}(M),$$

q.e.d. In particular, M is determined by the character α_M, so that the mapping $M \to \alpha_M$ is one-to-one.

If α is any character, then the linear functional $\int f(x)\overline{\alpha(x)} \, dx$ is multiplicative:

$$\int (f * g)(x)\overline{\alpha(x)} \, dx = \iint f(xy)g(y^{-1})\overline{\alpha(xy)}\overline{\alpha(y^{-1})} \, dx \, dy$$

$$= \iint f(x)\overline{\alpha(x)}g(y)\overline{\alpha(y)} \, dx \, dy$$

$$= \int f(x)\overline{\alpha(x)} \, dx \int g(y)\overline{\alpha(y)} \, dy.$$

It is therefore a homomorphism, and if M is its kernel the equation $\int f(x)\overline{\alpha(x)}\, dx = \int f(x)\overline{\alpha_M(x)}\, dx$ shows that $\alpha = \alpha_M$.

This theorem permits us to identify the set $\hat{G}$ of all the characters on G with the space $\mathfrak{M}$ of regular maximal ideals of $L^1(G)$ (which has already been identified with, and used interchangeably with, the subset of $(L^1)^*$ consisting of non-trivial homomorphisms). Generally from now on the domain of $\hat{f}$ will be taken to be $\hat{G}$. Also we shall use the notation (x, α) for the value of the character α at the group element x, so that the Fourier transform formula is now written:

$$\hat{f}(\alpha) = \int f(x)\overline{(x, \alpha)}\, dx.$$

34C. *The topology of $\hat{G}$ is that of uniform convergence on compact sets of G. That is, if C is a compact subset of G, $\epsilon > 0$ and $\alpha_0 \in \hat{G}$, then the set of characters $U(C, \epsilon, \alpha_0) = \{\alpha: |\, (x, \alpha) - (x, \alpha_0)\, | < \epsilon$ for all $x \in C\}$ is open, and the family of all such open sets is a basis for the topology of G.*

Proof. That $U(C, \epsilon, \alpha_0)$ is open follows at once from **5F**. Moreover the intersection of two such neighborhoods of α_0 includes a third: $U(C_1 \cup C_2, \min(\epsilon_1, \epsilon_2), \alpha_0) \subset U(C_1, \epsilon_1, \alpha_0) \cap U(C_2, \epsilon_2, \alpha_0)$. It is therefore sufficient to show that any neighborhood N of α_0 belonging to the usual sub-basis for the weak topology includes one of the above kind. Now N is of the form $N = \{\alpha: |\, \hat{f}(\alpha) - \hat{f}(\alpha_0)\, | < \delta\}$ for some $f \in L^1$ and $\delta > 0$. If C is chosen so that $\int_{C'} |\, f\, | < \delta/4$, $\epsilon = \delta/2\|\, f\, \|_1$ and $|\, (x, \alpha) - (x, \alpha_0)\, | < \epsilon$ on C, then $|\, \hat{f}(\alpha) - \hat{f}(\alpha_0)\, | \leqq \int |\, f(\bar{\alpha} - \bar{\alpha}_0)\, | \leqq \int_C + \int_{C'} \leqq \|\, f\, \|_1(\delta/2\|\, f\, \|_1) + 2(\delta/4) = \delta$. That is, $U(C, \epsilon, \alpha_0) \subset N$, q.e.d.

34D. Theorem. *The pointwise product of two characters is a character, and with this definition of multiplication $\hat{G}$ is a locally compact Abelian group.*

Proof. It is obvious that the product of two characters is a character. The constant 1 is a character and is the identity under multiplication. For any character α, the function α^{-1} is a character, and is the group inverse of α. These facts are evident, and there remains only the proof that multiplication is continuous. But if C is a compact subset of G, and $\epsilon > 0$, then

$$| \alpha\beta - \alpha_0\beta_0 | \leqq | \alpha - \alpha_0 | \cdot | \beta_0 | + | \alpha | \cdot | \beta - \beta_0 | < \epsilon$$

on C if $| \alpha - \alpha_0 | < \epsilon/2$ on C and $| \beta - \beta_0 | < \epsilon/2$ on C. It follows from **34C** that multiplication is continuous.

34E. *Remark:* It may seem to the reader that it would have been more natural to define α_M so that $\hat{f}(M) = \int f(x) \alpha_M(x) \, dx$ instead of $\hat{f}(M) = \int f(x) \overline{\alpha_M(x)} \, dx$. The only justification for our choice which can be given now is that it stays closer to the formalism of the scalar product and convolution; we have defined $\hat{f}(M)$ as $(f, \alpha_M) = [f * \alpha_M](e)$. Later we shall see that it leads to the conventional Fourier transform on the additive group of the reals, whereas the other choice leads to the conventional inverse Fourier transform. Notice that the character which we assign to a regular maximal ideal M is the inverse of the character assigned by the other formula.

§ 35. EXAMPLES

In calculating the character group for specific simple groups we have two possible methods of procedure. First, a character can be determined from the homomorphism of the group algebra associated with it, and this method was used in working out the examples of § 23. The second, and more natural, method in the present context, is to proceed directly from the definition of a character as a continuous homomorphism of G onto T. The direct verification of the correct topology for $\hat{G}$ takes perhaps a little longer because the theorem of **5G** on the equivalence of a weak topology with a given topology does not now intervene, but this is a minor objection at worst.

35A. Theorem. *The character group of a direct product $G_1 \times G_2$ of two locally compact Abelian groups is (isomorphic and homeomorphic to) the direct product $\hat{G}_1 \times \hat{G}_2$ of their character groups.*

Proof. The algebraic isomorphism is almost trivial. Any character α on $G_1 \times G_2$ is the product of its restrictions to G_1 and G_2, $\alpha(\langle x_1, x_2 \rangle) = \alpha(\langle x_1, e_2 \rangle)\alpha(\langle e_1, x_2 \rangle)$, and these restrictions are characters on $\hat{G}_1$ and $\hat{G}_2$ respectively. Conversely the function $\alpha(\langle x_1, x_2 \rangle) = \alpha_1(x_1)\alpha_2(x_2)$ defined as the product of two characters on G_1 and G_2 respectively is evidently a character on $G_1 \times G_2$. The one-to-one correspondence $\alpha \leftrightarrow \langle \alpha_1, \alpha_2 \rangle$ thus established between $(G_1 \times G_2)^\wedge$ and $\hat{G}_1 \times \hat{G}_2$ is clearly an isomorphism.

Let C be a compact subset of G containing $e = \langle e_1, e_2 \rangle$ and of the form $C = C_1 \times C_2$. If $|\alpha - \alpha^0| < \epsilon$ on C, then (taking $x_2 = e_2$) $|\alpha_1 - \alpha_1{}^0| < \epsilon$ on C_1 and, similarly, $|\alpha_2 - \alpha_2{}^0| < \epsilon$ on C_2. Conversely if $|\alpha_1 - \alpha_1{}^0| < \epsilon$ on C_1 and $|\alpha_2 - \alpha_2{}^0| < \epsilon$ on C_2, then $|\alpha - \alpha^0| = |\alpha_1\alpha_2 - \alpha_1{}^0\alpha_2{}^0| < 2\epsilon$ on C. It follows from **34C** that the topology of $(G_1 \times G_2)^\wedge$ is the Cartesian product topology of $\hat{G}_1 \times \hat{G}_2$. This fact also follows directly from the next theorem.

35B. Theorem. *If H is a closed subgroup of a locally compact Abelian group G, then the character group of the quotient group G/H is (isomorphic and homeomorphic to) the subgroup of $\hat{G}$ consisting of those characters on G which are constant on H and its cosets.*

Proof. If β is a character on G/H, then $\alpha(x) = \beta(Hx)$ is continuous on G and $\alpha(x_1 x_2) = \beta(Hx_1 x_2) = \beta(Hx_1 Hx_2) = \beta(Hx_1)\beta(Hx_2) = \alpha(x_1)\alpha(x_2)$, so that $\alpha(x)$ is a character on G. Conversely, if $\alpha(x)$ is a character on G which is constant on H, and therefore on each coset of H, the function β defined on G/H by $\beta(Hx) = \alpha(x)$ is a character.

We still have to show that the topology induced in this subgroup is the correct topology for the character group of G/H. We know that $C = \{Hx : x \in F\}$ is compact in G/H if F is a compact subset of G, and that every compact subset of G/H arises in this way (see **28C**). Then $|\beta(u) - \beta_0(u)| < \epsilon$ for all u in C if and only if $|\beta(Hx) - \beta_0(Hx)| < \epsilon$ for all x in F, and the result follows from **34C**.

35C. Theorem. *Every character $\alpha(x)$ on the additive group R of the real numbers is of the form $\alpha(x) = e^{iyx}$, and $\hat{R}$ is isomorphic and homeomorphic to R under the correspondence $e^{iyx} \leftrightarrow y$.*

Proof. Every continuous solution of the equation $\alpha(x + y) = \alpha(x)\alpha(y)$ is of the form $\alpha(x) \equiv e^{ax}$, and, since α is bounded, a must be pure imaginary. Thus $\alpha(x) = e^{iyx}$, for some real y. Conversely, every real number y defines the character e^{iyx}, and this one-to-one correspondence between R and $\hat{R}$ is an isomorphism by the law of exponents. The set of characters α such that $|\alpha(x) - 1| < \epsilon$ on $[-n, n]$ is a neighborhood of the identity character, and the set of such neighborhoods is a neighborhood basis around the identity. But $|e^{iyx} - 1| < \epsilon$ on $[-n, n]$ if and only if y is in the open interval $(-\delta, \delta)$, where $\delta = (2/n)$ arc sin $(\epsilon/2)$, so that the mapping $R \leftrightarrow \hat{R}$ is continuous both ways at the origin, and therefore everywhere.

35D. Theorem. *Each character on R/I is of the form $e^{2\pi inx}$, and $(R/I)^\wedge$ is isomorphic and homeomorphic to I under the correspondence $e^{2\pi inx} \leftrightarrow n$.*

Proof. This follows at once from **35B** and **C**; the character e^{iyx} has the constant value 1 on the integers if and only if $y = 2\pi m$ for some integer m, and the discrete subgroup $\{2\pi m\}$ of R is, of course, isomorphic (and trivially homeomorphic) to I.

35E. Theorem. *Each character on I is of the form $e^{2\pi ixn}$, where $0 \leq x < 1$, and $\hat{I}$ is isomorphic and homeomorphic to R/I under the correspondence $e^{2\pi ixn} \leftrightarrow x$.*

Proof. If $\alpha(n)$ is a character and $\alpha(1) = e^{2\pi iy}$, then $\alpha(n) = e^{2\pi iyn}$. It is clear that every such y defines a character, and that the mapping $y \to e^{2\pi iyn}$ is a homomorphism of R onto $\hat{I}$, with kernel I. Thus $\hat{I}$ is isomorphic to R/I. The set $\{y: 0 \leq y < 1$ and $|e^{2\pi iyn} - 1| < \epsilon, n = 1, \cdots, N\}$ is a basic neighborhood of e in I, and is easily seen to be an open interval about 0 in R/I. Conversely, any open interval about 0 in R/I is such a set (with $N = 1$ and ϵ to be determined), and the two topologies are identical.

35F. Each of the three groups R, R/I, and I is thus (isomorphic and homeomorphic to) its own second character group. These results are special cases of the Pontriagin duality theorem, which asserts that every locally compact Abelian group is its own second character group, and which we shall prove later in § 37.

The Fourier transform formula becomes for these groups the well-known formulas

$$\hat{f}(y) = \int_{-\infty}^{\infty} f(x)e^{-iyx}\, dx$$

$$\hat{f}(n) = C_n = \int_0^1 f(x)e^{-2\pi inx}\, dx$$

$$\hat{f}(x) = \int_I f(n)e^{-2\pi ixn}\, dn = \sum_{-\infty}^{\infty} C_n e^{-2\pi inx}.$$

The fact that the Fourier transform is initially a mapping of $L^1(G)$ into $\mathcal{C}(\hat{G})$ reduces here to the fact that each of the above integrals is absolutely convergent if $f \in L^1$ (which, in the last case, means $\sum_{-\infty}^{\infty} |C_n| < \infty$), and $\hat{f}$ is continuous and vanishes at infinity (in the first two cases). The corresponding formulas for multiple Fourier series, multiple Fourier transforms, and mixtures of these, can be written down by virtue of **35A**. The duality theorem can also be directly verified for these product groups.

§ 36. THE BOCHNER AND PLANCHEREL THEOREMS

36A. We shall now apply to the group algebra $L^1(G)$ the general commutative Banach algebra theory of § 25 and § 26, keeping in mind the extra information that the space $\hat{G}$ of characters (maximal ideal space of $L^1(G)$, space of homomorphisms) is itself a locally compact group, with its own Haar measure and its own group algebra.

We start with positivity. We have already seen in **31G** that a positive linear functional on $L^1(G)$ is extendable if and only if it is continuous. The Bochner theorem of **26I** can thus be taken to refer to positive definite functions $p \in L^\infty$. Moreover, the hypothesis in that theorem that $f^{*\wedge} = \hat{f}^-$ is now automatically met, for $f^{*\wedge}(\alpha) = \int \overline{f(x^{-1})}(x, \alpha)\, dx = \int \overline{f(x)}(x^{-1}, \alpha)\, dx = \left(\int f(x) \overline{(x, \alpha)\, dx}\right)^- = (\hat{f}(\alpha))^-$. Finally, if μ_p is the measure on $\hat{G}$ associated with the positive definite function $p \in L^\infty(G)$, we have

$$\int f(x)\overline{p(x)}\, dx = \int \hat{f}(\alpha)\, d\mu_p(\alpha) = \int f(x)\left[\int \overline{(x, \alpha)}\, d\mu_p(\alpha)\right] dx$$

for all $f \in L^1$, implying that $p(x) = \int (x, \alpha) \, d\mu_p(\alpha)$ for almost all x. Any function p defined this way is positive definite, for it clearly belongs to L^∞, and

$$(f * f^*, p) = \iint (f * f^*)(x)\overline{(x, \alpha)} \, dx \, d\mu_p(\alpha) = \int |\hat{f}|^2 \, d\mu_p \geqq 0.$$

The Bochner theorem can therefore be reformulated in the present context as follows:

Theorem. *The formula*

$$p(x) = \int (x, \alpha) \, d\mu(\alpha)$$

sets up a norm preserving isomorphism between the convex set of all finite positive Baire measures μ on $\hat{G}$ and the convex set of all positive definite functions $p \in L^\infty(G)$.

Corollary. *Every positive definite function $p \in L^\infty(G)$ is (essentially) uniformly continuous.*

Proof. This is because the measure μ_p is mostly confined to a compact set C and the character function (x, α) is continuous in x uniformly over all $\alpha \in C$. Given ϵ we choose a compact subset C of $\hat{G}$ such that $\mu_p(C') < \epsilon/4$ (C' being the complement of C), and then find a neighborhood V of the identity in G such that $|(x_1, \alpha) - (x_2, \alpha)| < \epsilon/2\mu_p(C)$ if $x_1 x_2^{-1} \in V$ and $\alpha \in C$ (by **5F** again). Then

$$\left| \int [(x_1, \alpha) - (x_2, \alpha)] \, d\mu_p(\alpha) \right| \leqq \int_C + \int_{C'} \leqq \frac{\epsilon}{2} + \frac{\epsilon}{2}$$

when $x_1 x_2^{-1} \in V$. Thus $\int (x, \alpha) \, d\mu_p(\alpha)$ is uniformly continuous.

36B. Let $\mathcal{P}$ be the class of positive definite functions. Because $L^1(G)$ has an approximate identity, it follows that *the vector space $[L^1 \cap \mathcal{P}]$ generated by $L^1 \cap \mathcal{P}$ is dense in L^1.* For $L^1 \cap L^\infty$ is dense in L^1, and if $f \in L^1 \cap L^\infty$ and u runs through an approximate identity (also chosen from $L^1 \cap L^\infty$), then $f * u$ is at once an approximation to f and a linear combination of four positive definite functions like $(f + u) * (f + u)^*$. By exactly the same argument $[L^1 \cap \mathcal{P}]$ is dense in $L^2(G)$.

The Bochner theorem above and the abstract Plancherel-like theorem of **26J** now lead us to the following L^1-inversion theorem.

Theorem. *If $f \in [L^1 \cap \mathcal{O}]$, then $\hat{f} \in \hat{L}^1$ and*

$$f(x) = \int (x, \alpha) \hat{f}(\alpha) \, d\alpha$$

for almost all x, where $d\alpha$ is the Haar measure of $\hat{G}$ suitably normalized.

Proof. Let φ be the positive functional defined on the ideal L^0 of uniformly continuous functions of L^1 by $\varphi(f) = f(e)$. We have observed earlier (**31G**) that a function $p \in L^1 \cap \mathcal{O}$ is positive definite with respect to φ in the sense of the Plancherel-type theorem of **26J**, and we can therefore conclude from that theorem that there exists a unique positive Baire measure m on $\hat{G}$ such that $\hat{p} \in L^1(m)$ and $(f, p) = \int f \hat{p} \, dm$ whenever $p \in L^1 \cap \mathcal{O}$ and $f \in L^1$. The formula $p(x) = \int (x, \alpha) \hat{p}(\alpha) \, dm(\alpha)$ is the same as that of the Bochner theorem in **36A**. We have left only to prove that m is the Haar measure of $\hat{G}$. But if $p \in L^1 \cap \mathcal{O}$, then a direct check shows that $p(x)\overline{(x, \alpha_0)} \in L^1 \cap \mathcal{O}$ for any character α_0, and we have

$$\int \hat{p}(\alpha) \, dm(\alpha) = p(e) = p(e)(e, \alpha_0) = \int \hat{p}(\alpha_0 \alpha) \, dm(\alpha),$$

where we have used the directly verifiable fact that $\hat{p}_{\alpha_0}$ is the Fourier transform of $p\bar{\alpha}_0$. Since the algebra generated by $L^1 \cap \mathcal{O}$ is dense in L^1, and the transforms $\hat{p}$ are therefore dense in $\mathcal{C}(\hat{G})$, the above equation shows m to be translation invariant and therefore the Haar measure of $\hat{G}$.

Remark: If T is the Fourier transformation $f \to \hat{f}$ from $L^1(G)$ into $\mathcal{C}(\hat{G})$, then the mapping $\mu \to p$ defined by $p(x) = \int (x, \alpha) \, d\mu(\alpha)$ from the space of all bounded complex-valued Baire measures on $\hat{G}$ ($\mathcal{C}(\hat{G})^*$) into $L^\infty(G)$ ($= L^1(G)^*$) is the adjoint mapping T^*. It must be remembered, however, that the identification of L^∞ with $(L^1)^*$ is taken to be a *conjugate-linear* mapping in order that the ordinary scalar product formula can be used: $P(f) = \int f(x)\overline{p(x)}\, dx$, where P and p are corresponding elements of $(L^1)^*$ and L^∞. The same remark holds for the identification of $\mathcal{C}(\hat{G})^*$ with the space of bounded Baire measures. It is this twist which is responsible for the factor (x, α) in the integrand of T^* whereas $\overline{(x, \alpha)}$ occurs in that of T.

The inversion theorem above shows that $T^* = T^{-1}$ when T is restricted to $[L^1 \cap \mathcal{P}]$. However the function $\hat{f}$ occurring in the range of T is taken as an element of $\mathcal{C}(\hat{G})$, whereas the same function as an element of the domain of T^* is considered as a measure $(\hat{f}(\alpha)\, d\alpha)$ on $\hat{G}$. This discrepancy vanishes when the L^2 norms are used; T then turns out to be a unitary mapping and the equation $T^* = T^{-1}$ is proper. This is the point of the Plancherel theorem, both in its general form (**26J**) and its group form (**36D**).

36C. It is evident that the inversion formula must be multiplied by a constant if a different Haar measure is used on G; that is, the inversion formula picks out a unique Haar measure on $\hat{G}$ in terms of the given Haar measure on G. In specific cases the correct determination of m on $\hat{G}$ in terms of μ on G is an interesting and non-trivial problem. A possible procedure is to calculate explicitly the transform and inverse transform of some particular, easily handled function.

By way of illustration consider the group R and the function $f(x) = e^{-x^2/2}$. If in the formula

$$\hat{f}(y) = \int_{-\infty}^{\infty} e^{-x^2/2} e^{-iyx}\, dx$$

we differentiate with respect to y and then integrate by parts we find that $u = \hat{f}(y)$ satisfies the differential equation $du = -uy\, dy$, so that $\hat{f}(y) = Ce^{-y^2/2}$. In order to determine C we observe that $\hat{f}(0) = C = \int_{-\infty}^{\infty} e^{-x^2/2}\, dx$. Then

$$C^2 = \int_{-\infty}^{\infty}\int_{-\infty}^{\infty} e^{-(x^2 + y^2)/2}\, dx\, dy = \int_{0}^{2\pi}\int_{0}^{\infty} e^{-r^2/2} r\, dr\, d\theta = 2\pi,$$

and $C = (2\pi)^{\frac{1}{2}}$. Therefore if we take Haar measure on R to be $dx/(2\pi)^{\frac{1}{2}}$ instead of dx the function $e^{-x^2/2}$ transforms into itself.

Now we must choose Haar measure $C\,dy$ on $\hat{R} = R$ so that the inverse formula

$$e^{-x^2/2} = \int_{-\infty}^{\infty} e^{-v^2/2} e^{ivx} C\,dy$$

holds. Since this equation is real it is the same as its complex conjugate above and C must be $1/(2\pi)^{1/2}$. Thus if $f(x) \in [L^1 \cap P]$, then $\hat{f}(y) \in [L^1 \cap P]$, where

and

$$\hat{f}(y) = \frac{1}{(2\pi)^{1/2}} \int_{-\infty}^{\infty} f(x) e^{-ivx}\,dx$$

$$f(x) = \frac{1}{(2\pi)^{1/2}} \int_{-\infty}^{\infty} \hat{f}(y) e^{ivx}\,dy;$$

both integrals being absolutely convergent. This, of course, is the classical choice for the Haar measures on R and $\hat{R} = R$. It is easy to see that all other pairs of associated measures are of the form $(\lambda\,dx/(2\pi)^{1/2}, dy/\lambda(2\pi)^{1/2})$.

The proper pairing of measures for compact and discrete groups such as R/I and I will follow from **38B**.

36D. The Plancherel theorem. *The Fourier transformation $f \to Tf = \hat{f}$ preserves scalar products when confined to $[L^1 \cap \mathcal{P}]$, and its L^2-closure is a unitary mapping of $L^2(G)$ onto $L^2(\hat{G})$.*

Proof. The heart of the Plancherel theorem has already been established in **26J** and restated implicitly in the L^1 inversion theorem above, namely, that the restriction to $[L^1 \cap \mathcal{P}]$ of the Fourier transformation $f \to Tf = \hat{f}$ satisfies $(Tf, Tp) = (\hat{f}, \hat{p}) = \varphi(f * p) = (f, p)$ and is therefore an *isometry* when the L^2 norms are used.

The remainder of the theorem is that T has a unique extension to a unitary mapping of the whole of $L^2(G)$ onto the whole of $L^2(\hat{G})$. That $[L^1 \cap \mathcal{P}]$ is dense in L^2 has been remarked earlier; it depends on the fact that convolutions $f * g$ with $f, g \in L^1 \cap L^2$ belong to $[L^1 \cap \mathcal{P}]$ and are dense in L^2. Thus T has a unique extension to the whole of $L^2(G)$, and the extended range is a complete (and hence closed) subspace of $L^2(\hat{G})$. We will be finished therefore if we can prove that the range of T is dense in $L^2(\hat{G})$.

Now if $F \in L^1 \cap L^2(\hat{G})$, then $f = T^*(F) \in L^2(G)$. For f is bounded and uniformly continuous, and if $g \in L^1 \cap L^2(G)$, then $|(g,f)| = |(\hat{g}, F)| \leqq \|F\|_2 \|\hat{g}\|_2 = \|F\|_2 \cdot \|g\|_2$, proving that $f \in L^2(G)$ and $\|f\|_2 \leqq \|F\|_2$. If also $H \in L^1 \cap L^2(\hat{G})$, then $h \in L^2(G)$ and $hf = T^*(H * F) \in [L^1 \cap \mathcal{P}]$. Thus $H * F = T(hf)$ by the inversion theorem, and since such convolutions $H * F$ are dense in $L^2(\hat{G})$ we are finished.

We have also, as a corollary of the above method of proof, the following:

Corollary. $L^1(G)$ *is semi-simple and regular.*

Proof. If $f \in L^1$ and $g \in L^2$, then $f * g \in L^2$ and $T(f * g) = \hat{f}\hat{g}$. If $f \neq 0$, the convolution operator U_f defined on $L^2(G)$ by f is not zero (**32C**) and therefore the operator on $L^2(\hat{G})$ defined by multiplication by $\hat{f}$ is not zero. That is, if $f \neq 0$ in $L^1(G)$, then $\hat{f} \neq 0$ in $\mathcal{C}(\hat{G})$, and $L^1(G)$ is semi-simple.

The reader is reminded that $L^1(G)$ is regular if for every closed set $F \subset \hat{G}$ and every point $\alpha \notin F$ there exists $f \in L^1$ such that $\hat{f} = 0$ on F and $\hat{f}(\alpha) \neq 0$. We prove directly the stronger assertion about local identities in L^1 (**25C**), namely, that if F and U are subsets of $\hat{G}$ such that F is compact, U is open and $F \subset U$, then there exists $f \in L^1(G)$ such that $\hat{f} \equiv 1$ on F and $f \equiv 0$ on U'. To see this we choose a symmetric Baire neighborhood V of the identity in $\hat{G}$ such that $V^3 F \subset U$ and let $\hat{g}$ and $\hat{h}$ be respectively the characteristic functions of V and a Baire open set between VF and V^2F. Then the convolution $\hat{g} * \hat{h}$ is identically equal to $m(V)$ on F and to zero outside of U. Since $\hat{g} * \hat{h} = T(gh)$, the desired function is $f = gh/m(V)$.

36E. We conclude this section by translating the representation theorem of **26F** to the group setting. If T ($s \to T_s$) is any unitary representation of an Abelian group by unitary transformations on a Hilbert space H, the considerations of § 32 tell us that T is completely equivalent to a *-representation of the group algebra $L^1(G)$. Then, as in **26F**, T can be "transferred" to the algebra of transforms $\hat{f}$ and then extended to a *-representation of the algebra of all bounded Baire functions on $\hat{G}$. The trilinear functional $I(\hat{f}, x, y) = (T_f x, y)$, for fixed x and y, is a bounded complex-valued integral on $\mathcal{C}(\hat{G})$, and satisfies $I(F\hat{g}, x, y) =$

$I(F, T_gx, y)$ for $F \subset \mathcal{C}(\hat{G})$ and $g \in L^1(G)$. Thus $(T_sT_gx, y) =$ $(T_{g_s-1}x, y) = I((s^{-1}, \alpha)\hat{g}(\alpha), x, y) = I(\bar{s}, T_gx, y)$. The operator T_{φ_E} corresponding to the characteristic function φ_E of the set E is a projection, designated P_E, and (P_Ex, y) is the complex-valued measure corresponding to the above integral. (The correspondence $E \rightarrow P_E$ is called a projection-valued measure.) The above equation can therefore be written

$$(T_su, y) = \int \overline{(s, \alpha)} \, d(P_\alpha u, y),$$

where we have set $u = T_gx$. Since such elements are dense in H, this restriction on u can be dropped. We can summarize this result, using the symbolic integral of spectral theory, as follows:

Stone's theorem. *If T is a unitary representation of the locally compact Abelian group G by unitary transformations on a Hilbert space H, then there exists a projection-valued measure P_E on $\hat{G}$ such that*

$$T_s = \int \overline{(s, \alpha)} \, dP_\alpha.$$

The projections P_E form the so-called spectral family of the commutative family of operators T_s. Each such projection defines a reduction of the representation T into the direct sum of TP_E operating on the range of P_E and $T(1 - P_E) = TP_{E'}$ operating on its orthogonal complement in H. The further analysis of the reducibility of T depends directly on the multiplicity theory of the family of projections P_E and is beyond the reach of this book.

§ 37. MISCELLANEOUS THEOREMS

This section contains the Tauberian theorem and its generalizations, the Pontriagin duality theorem, and a simple case of the Poisson summation formula.

37A. The Tauberian theorem, as proved in **25D**, applies to any regular commutative Banach algebra, and hence (see **36D**, corollary) to the L^1-group algebra of an Abelian locally compact group. The extra condition of the theorem, that the elements whose Fourier transforms vanish off compact sets are dense in the algebra, is always satisfied in this case. For $\hat{L}$ is dense in

$L^{2\wedge}$ and hence the elements of L^2 whose transforms lie in $\hat{L}$ are dense in L^2. Since every function of L^1 is a product of functions in L^2, it follows via the Schwarz inequality that the functions of L^1 whose transforms lie in $\hat{L}$ are dense in L^1. We can now restate the Tauberian theorem.

Theorem. *If G is a locally compact Abelian group, then every proper closed ideal of $L^1(G)$ is included in a regular maximal ideal.*

Corollary 1. *If $f \in L^1$ is such that $\hat{f}$ never vanishes, then the translates of f generate L^1.*

Proof. By hypothesis f lies in no regular maximal ideal. The closed subspace generated by the translates of f is (by **31F**) a closed ideal, and therefore by the theorem is the whole of L^1.

Corollary 2. (The generalized Wiener Tauberian theorem.) *Let G be any locally compact Abelian group which is not compact. Let $k \in L^1$ be such that $\hat{k}$ never vanishes and let $g \in L^\infty$ be such that the continuous function $k * g$ vanishes at infinity. Then $f * g$ vanishes at infinity for every $f \in L^1$.*

Proof. The set I of functions $f \in L^1$ such that $f * g$ vanishes at infinity is clearly a linear subspace, and I is invariant since $f_x * g = (f * g)_x$ vanishes at infinity if $f * g$ does.

I is also closed. For if $f \in \bar{I}$ and ϵ is given, we can choose $h \in I$ such that $\| f - h \|_1 < \epsilon/(2\| g \|_\infty)$, and we can choose a compact set C such that $| (h * g)(x) | < \epsilon/2$ outside of C. Since $| (f * g)(x) - (h * g)(x) | \leqq \| f - h \|_1 \| g \|_\infty < \epsilon/2$, it follows that $| (f * g)(x) | < \epsilon$ outside of C. Therefore $f * g$ vanishes at infinity and $f \in I$.

Therefore (by **31F** again) I is a closed ideal, and since it contains k it is included in no regular maximal ideal. It follows from the theorem that $I = L^1$.

Corollary 3. (The Wiener Tauberian theorem.) *If $k \in L^1(R)$ is such that $\hat{k}$ never vanishes and $g \in L^\infty$ is such that $(k * g)(x) \to 0$ as $x \to +\infty$, then $(f * g)(x) \to 0$ as $x \to +\infty$ for every $f \in L^1$.*

This is not quite a special case of the above corollary since it concerns being zero at infinity only in one direction. However the proof is the same.

37B. The Tauberian theorem in the general form of the theorem of **37A** asserts that a closed ideal is the kernel of its hull if its hull is empty, and is thus a positive solution to the general problem of determining under what conditions a closed ideal in a group algebra $L^1(G)$ is equal to the kernel of its hull. That the answer can be negative was shown by Schwartz (C. R. Acad. Sci. Paris 227, 424–426 (1948)), who gave a counter example in the case where G is the additive group of Euclidean 3-space. The positive theorem also holds for one-point hulls. This was proved for the real line by Segal [44] and extended, by means of structure theory, to general locally compact Abelian groups by Kaplansky [27]. Helson has given a proof of the generalized theorem which is independent of structural considerations [24]. In the proof below we use a function which has been discussed both by Helson and by Reiter (Reiter's paper is unpublished at the time of writing) to give a direct proof that the condition of Ditkin (**25F**) is valid in the L^1 algebra of any locally compact Abelian group. This immediately implies (by **25F**) the strongest known positive theorem.

Lemma. *If a regular, semi-simple Banach algebra A has an approximate identity and has the property (of Wiener's theorem) that elements x such that $\hat{x}$ has compact support are dense in A, then $\hat{A}$ satisfies the condition D at infinity.*

Proof. This lemma has content only if $\mathfrak{M}$ is not compact. We must show that for every x and ϵ there exists y such that $\hat{y}$ has compact support and $\| xy - x \| < \epsilon$. Because A has an approximate identity there exists u such that $\| xu - x \| < \epsilon/2$, and then, by the Wiener condition, there exists y such that $\hat{y}$ has compact support and $\| y - u \| < \epsilon/2\| x \|$. Then $\| xy - xu \| < \epsilon/2$ and $\| xy - x \| < \epsilon$, q.e.d.

Corollary. *It then follows that, if I is a closed ideal with a compact hull, then I contains every element x such that hull $(I) \subset int$ $(hull\ (x))$.*

Proof. It follows from the theorem that there exists y such that $\hat{y}$ has compact support and $\| xy - x \| < \epsilon$. But then $xy \in I$ by **25D**, and since I is closed we have $x \in I$.

37C. In order to establish the condition D at finite points we introduce the function mentioned above.

Let U be any symmetric Baire neighborhood of the identity $\hat{e}$ in $\hat{G}$ such that $m(U) < \infty$, where m is the Haar measure of $\hat{G}$. Let V be any second such neighborhood whose closure is compact and included in U, and taken large enough in U so that $m(U)/m(V) < 2$. Let $\hat{u}$ and $\hat{v}$ be the characteristic functions of U and V respectively, and let u and v be their inverse Fourier transforms in $L^2(G)$. Then the function $j = uv/m(V)$ belongs to $L^1(G)$, and $\|j\|_1 \leqq \|u\|_2 \|v\|_2/m(V) = \|\hat{u}\|_2 \|\hat{v}\|_2/m(V) = [m(U)/m(V)]^{\frac{1}{2}} < 2$. Moreover if W is a neighborhood of $\hat{e}$ such that $VW \subset U$ then $\hat{j} = \hat{u} * \hat{v}/m(V) = (1/m(V)) \int \hat{u}(\alpha)\hat{v}(\alpha^{-1}\beta)\,d\alpha$ $= 1$ if $\beta \in W$.

Lemma. *Given any compact set $C \subset G$ and any $\epsilon > 0$ there exists a function $j \in L^1$ such that $\hat{j} \equiv 1$ in some neighborhood of the identity $\hat{e}$ in $\hat{G}$, $\|j\|_1 < 2$, and $\|j - j_x\|_1 < \epsilon$ for every $x \in C$.*

Proof. We define j as above, taking for U any symmetric Baire open subset of the open set $\{\alpha : |1 - (x, \alpha)| < \epsilon/4$ for all $x \in C\}$ (see **5F**). Since $j - j_x = [u(v - v_x) + v_x(u - u_x)]/m(V)$ and since, with the above choice of U, $\|u - u_x\|_2^2 = \int |\hat{u}(\alpha)(1 - (x, \alpha))|^2\,d\alpha \leqq m(U) \operatorname{lub}_{\alpha \in U} |1 - (x, \alpha)|^2 < m(U)(\epsilon/4)^2$ if $x \in C$ (and similarly for $\|v - v_x\|_2$), we have $\|j - j_x\|_1 \leqq 2[m(U)m(V)]^{\frac{1}{2}}\epsilon/4m(V) < \epsilon$ for every $x \in C$, q.e.d.

Corollary. *If $f \in L^1(G)$ and $\hat{f}(\hat{e}) = 0$ then $f * j \to 0$ as U decreases through the symmetric Baire neighborhoods of $\hat{e}$.*

Proof. Given δ we choose C in the above lemma to be symmetric and satisfying $\int_{C'} |f| < \delta/8$, and set $\epsilon = \delta/2\|f\|_1$. Then $f * j(y) = \int f(x)j(x^{-1}y)\,dx = \int f(x)(j(x^{-1}y) - j(y))\,dx$, and $\|f * j\|_1 \leqq \int |f(x)| \|j_{x^{-1}} - j\|_1\,dx = \int_C + \int_{C'} < \|f\|_1\epsilon + (\delta/8)4 = \delta$.

This corollary leads at once to the condition D at the identity $\hat{e}$; it then follows for other points upon translating.

Theorem. *There exists a uniformly bounded directed set of functions $v \in L^1(G)$ such that $\hat{v} \equiv 0$ in a neighborhood of $\hat{e}$ and such that $f * v \to f$ for every $f \in L^1(G)$ such that $\hat{f}(\hat{e}) = 0$.*

Proof. Let u run through an approximate identity and set $v = u - j * u$. Then $\|v\|_1 \leqq 3$ and $\hat{v} = \hat{u} - \hat{u}\hat{j} = 0$ in the neighborhood of $\hat{e}$ where $\hat{j} = 1$. Also $\|f - f * v\|_1 \leqq \|f - f * u\|_1 + \|f * j\|_1 \| u\|_1 \to 0$, q.e.d.

It follows that Silov's theorem (**25F**) is valid for the group algebras of locally compact Abelian groups. We restate the theorem slightly.

Theorem. *Let I be a closed ideal in $L^1(G)$ and f a function of $L^1(G)$ such that $\hat{f}(\alpha) = 0$ whenever α is a character at which every function of I vanishes; that is, hull $(I) \subset$ hull (f), or $f \in k(h(I))$. Suppose furthermore that the part of the boundary of hull (f) which is included in hull (I) includes no non-zero perfect set. Then $f \in I$.*

Corollary. *If I is a closed ideal in $L^1(G)$ whose hull is discrete (i.e., consists entirely of isolated points) then $I = k(h(I))$.*

37D. We have seen that the character function (x, α), $x \in G$, $\alpha \in \hat{G}$, is continuous in x and α together. Also that $(x, \alpha_1\alpha_2) = (x, \alpha_1)(x, \alpha_2)$. Therefore, every element x in G defines a character on $\hat{G}$, and the mapping of G into $\hat{\hat{G}}$ thus defined is clearly an algebraic homomorphism. The Pontriagin duality theorem asserts that this mapping is an isomorphism and a homeomorphism onto $\hat{\hat{G}}$, so that G can be identified with $\hat{\hat{G}}$.

By using the considerable analytic apparatus available to us we can deduce a short proof of this theorem.

The space $[L^1 \cap \mathcal{O}]$ corresponds exactly to the space $[\hat{L}^1 \cap \hat{\mathcal{O}}]$ under the Fourier transform and forms an algebra A of uniformly continuous functions vanishing at infinity and separating the points of G (since it includes the subalgebra generated by $[(L^1 \cap L^2) * (L^1 \cap L^2)]$; see **36A, B**). The weak topology determined on G by A is therefore the given topology of G (**5G**). But, as the Fourier transform of a function $\hat{f} \in [\hat{L}^1 \cap \hat{\mathcal{O}}]$, each such f has a unique extension to the whole of $\hat{G}$, and under this extension $[L^1 \cap \mathcal{O}]$ is isomorphic to $[L^1 \cap \mathcal{O}](\hat{G})$. Since the topology of $\hat{G}$ is the weak topology defined by these extended functions, it follows that G is imbedded in $\hat{\hat{G}}$ with the relative topology it acquires from $\hat{\hat{G}}$, and that G is dense in $\hat{\hat{G}}$. (Otherwise we could construct a non-zero convolution of two functions of $(L^1 \cap L^2)(\hat{G})$ which vanishes identically on G, contradicting the isomorphic extension.) Since the subsets of G and $\hat{\hat{G}}$ on which $|\hat{\hat{f}}| > \epsilon > 0$ have compact closures in both topologies and one is dense in the other, we can conclude that their closures are identical. Therefore $G = \hat{\hat{G}}$, finishing the proof of the duality theorem.

37E. The Poisson summation formula. Let f be a function on $R = (-\infty, \infty)$ and let F be its Fourier transform. Let α be any positive number and let $\beta = 2\pi/\alpha$. Then under suitable conditions on f it can be shown that

$$\sqrt{\alpha} \sum_{-\infty}^{\infty} f(n\alpha) = \sqrt{\beta} \sum_{-\infty}^{\infty} F(n\beta),$$

which is the Poisson formula.

The formal proof proceeds as follows. The function $g(x) = \sum_{-\infty}^{\infty} f(x + n\alpha)$ is periodic and hence can be considered to be defined on the reals mod α. The characters of this quotient group are the functions $e^{imx(2\pi/\alpha)} = e^{im\beta x}$, and its character group, as a subgroup of $\hat{R} = R$ (see **35B**), is the discrete group $\{m\beta\}_{-\infty}^{\infty}$. Then

$$G(m\beta) = \frac{1}{\alpha} \int_0^{\alpha} e^{-im\beta x} g(x)\, dx = \frac{1}{\alpha} \sum_{n=-\infty}^{\infty} \int_0^{\alpha} e^{-im\beta x} f(x + n\alpha)\, dx$$

$$= \frac{1}{\alpha} \sum_{n=-\infty}^{\infty} \int_{n\alpha}^{(n+1)\alpha} e^{-im\beta x} f(x)\, dx = \frac{1}{\alpha} \int_{-\infty}^{\infty} e^{-im\beta x} f(x)\, dx$$

$$= \frac{\sqrt{2\pi}}{\alpha} F(m\beta) = \sqrt{\frac{\beta}{\alpha}}\, F(m\beta).$$

But by the inversion theorem $g(x) = \sum_{-\infty}^{\infty} e^{im\beta x} G(m\beta)$. Therefore

$$\sum_{-\infty}^{\infty} f(n\alpha) = g(0) = \sum_{-\infty}^{\infty} G(m\beta) = \sqrt{\frac{\beta}{\alpha}} \sum_{-\infty}^{\infty} F(m\beta),$$

giving the Poisson formula.

We now present a proof in the general situation, with the function f severely enough restricted so that the steps of the formal proof are obviously valid. More could be proved by careful arguing, but our purpose here is only to display the group setting of the theorem.

Theorem. *Let G be a locally compact Abelian group and let H be a closed subgroup. Let the Haar measures on G, H and G/H be adjusted so that* $\int_G = \int_{G/H}\int_H$, *and let f be a function of* $[L^1 \cap \mathcal{P}](G)$ *such that* $g(y) = \int_H f(xy)\, dx$ *is a continuous function (on G/H) of y. Then*

$$\int_H f(x)\, dx = \int_{\widehat{G/H}} \hat{f}(\alpha)\, d\alpha.$$

Proof. The character group of G/H is by **35B** the set of characters $\alpha \in \hat{G}$ such that $(xy, \alpha) = (y, \alpha)$ for every $x \in H$. Then

$$\hat{g}(\alpha) = \int_{G/H} \overline{(y, \alpha)}g(y)\, dy = \int_{G/H} \overline{(y, \alpha)} \left[\int_H f(xy)\, dx \right] dy$$

$$\int_{G/H}\int_H \overline{(xy, \alpha)} f(xy)\, dx\, dy = \int_G \overline{(x, \alpha)}f(x)\, dx = \hat{f}(\alpha).$$

But by the inversion theorem $g(y) = \int_{\widehat{G/H}} (x, \alpha)\hat{g}(\alpha)\, d\alpha$. Therefore

$$\int_H f(x)\, dx = g(0) = \int_{\widehat{G/H}} \hat{g}(\alpha)\, d\alpha = \int_{\widehat{G/H}} \hat{f}(\alpha)\, d\alpha,$$

which is the desired formula.

§ 38. COMPACT ABELIAN GROUPS AND GENERALIZED FOURIER SERIES

If G is compact and Abelian, its Fourier transform theory is contained in the analysis of H^*-algebras carried out in § 27 and its application in Chapter VIII. However, a simple direct discussion will be given here.

38A. Theorem. *G is compact if and only if $\hat{G}$ is discrete.*

Proof. Reversing the roles of G and $\hat{G}$ we remark, first, that, if G is discrete, then $L^1(G)$ has an identity (**31D**) and $\hat{G}$ is compact as the maximal ideal space of a commutative Banach algebra with an identity (**19B**). Second, since $L^1(G)$ is semi-simple and self-adjoint, it follows from **26B** that, if $\hat{G}$ is compact, then an identity can actually be constructed for $L^1(G)$ by applying the analytic function 1 to an everywhere positive function $\hat{f}$ in the transform algebra, and (by **31D** again) G is discrete. The theorem as stated now follows from the duality theorem.

The second part of the argument can be proved directly as follows. If G is compact, then the set of characters α such that $\| \alpha - 1 \|_\infty < \frac{1}{2}$ is an open neighborhood of the identity character 1 (by **34C**) which obviously contains only the identity character; thus the topology of $\hat{G}$ is discrete.

38B. Theorem. *If G is compact and its Haar measure is normalized so that $\mu(G) = 1$, then the inversion theorem requires that the Haar measure on $\hat{G}$ be normalized so that the measure of each point is* 1.

Proof. If the measure of a point in $\hat{G}$ is 1, then the identity of $L^1(\hat{G})$ is the function $\hat{u}$ which has the value 1 at $\alpha = \hat{e}$ and the value 0 elsewhere. Since convolutions on $\hat{G}$ correspond to ordinary pointwise products on G, $\hat{u}$ is the transform of $u \equiv 1$. Thus $\hat{u}(\alpha) = \int \overline{(x, \alpha)} \, dx = 1$ if $\alpha = \hat{e}$ and $= 0$ otherwise. Since $(x, \hat{e}) \equiv 1$, this implies that $\mu(G) = \int 1 \, dx = 1$, so that the measures match as stated in the theorem.

The above identity can also be written $\int (x, \alpha_1)\overline{(x, \alpha_2)} \, dx = \int \overline{(x, \alpha_2\alpha_1{}^{-1})} \, dx = 1$ if $\alpha_2 = \alpha_1$ and $= 0$ if $\alpha_2 \neq \alpha_1$. Thus:

Corollary. *The characters on G form an orthonormal set.*

38C. *The characters form a complete orthonormal set in $L^2(G)$ and the (Fourier series) development $f(x) = \sum_1^\infty (f, \alpha_n)\alpha_n$ of a function $f \in L^2(G)$ is the inverse Fourier transform.*

Proof. If $f(x) \in L^2$, then $\hat{f} \in \hat{L}^2$, so that $\hat{f}(\alpha) = 0$ except on a countable set $\{\alpha_n\}$ and, if $c_n = \hat{f}(\alpha_n) = \int f(x)\overline{(x, \alpha_n)} \, dx = (f, \alpha_n)$, then

$$\int | f |^2 \, dx = \int | \hat{f} |^2 \, d\alpha = \sum_1^\infty | c_n |^2,$$

which is Parseval's equation. The continuous function $f_n = \sum_1^n c_i \alpha_i$ is by its definition the inverse Fourier transform of the finite-valued function $\hat{f}_n$ equal to c_i at α_i if $i \leqq n$ and equal to 0 elsewhere. Since $\hat{f}_n$ clearly converges (L^2) to $\hat{f}$, it follows that f_n converges (L^2) to f and that the formula $f(x) = \sum_1^\infty c_n(x, \alpha_n) = \int \hat{f}(\alpha)(x, \alpha) \, d\alpha$ is the inverse Fourier transform.

38D. *Every continuous function on G can be uniformly approximated by finite linear combinations of characters.*

This follows from the Stone-Weierstrass theorem, for the characters include the constant 1 and separate points. Since the product of two characters is a character, the set of finite linear combinations of characters forms an algebra having all the properties required.

A direct proof can be given. First f is approximated uniformly by $f * u$ where u is an element of an approximate identity in $L^1(G)$. Then, as above, f is approximated L^2 by $f_n = \sum_1^n c_i(x, \alpha_i)$ and u by $u_n = \sum_1^n d_i(x, \alpha_i)$, where the sequence $\{\alpha_n\}$ includes all the characters at which either $\hat{f}$ or $\hat{u} \neq 0$. Then $f * u$ is approximated uniformly by $f_n * u_n = \sum_1^n c_i d_i(x, \alpha_i)$. The last equation can be verified by writing out the convolution or by remembering that $f_n * u_n = T^*(\hat{f}_n \hat{g}_n)$.

The partial sums of the Fourier expansion of f will in general not give uniform approximations.

Chapter VIII

COMPACT GROUPS AND ALMOST PERIODIC FUNCTIONS

If G is compact $L^2(G)$ is an H^*-algebra, and the whole theory of § 27 can be taken over. Thus $L^2(G)$ is the direct sum of its (mutually orthogonal) minimal two-sided ideals, and it turns out that these are all finite dimensional and consist entirely of continuous functions; their identities (generating idempotents) are the *characters* of G. These facts, which are in direct generalization of the classical theory of the group algebra of a finite group, are presented in § 39, and in § 40 they are used to obtain the complete structure of unitary representations of G. § 41 presents the theory of almost periodic functions on a general group, following a new approach.

§ 39. THE GROUP ALGEBRA OF A COMPACT GROUP

39A. The Haar measure of a compact group is finite and generally normalized to be 1. Then $\| f \|_1 \leq \| f \|_2 \leq \| f \|_\infty$. Moreover, on any unimodular group the Schwarz inequality implies that $\| f * g \|_\infty \leq \| f \|_2 \| g \|_2$. Together, these two inequalities show that L^∞ and L^2 are both Banach algebras. The latter is in fact an H^*-algebra; it is already a Hilbert space, and the further requirements that $(f * g, h) = (g, f^* * h)$, $\| f^* \|_2 = \| f \|_2$ and $f_* f^* \neq 0$ if $f \neq 0$ are all directly verifiable. Thus $f * f^*$ is continuous and $(f * f^*)(e) = \| f \|_2{}^2 > 0$ if $f \neq 0$, and the other two conditions follow from the inverse invariance of the Haar measure of G. The fact that $L^2(G)$ is an H^*-algebra means that its algebraic structure is known in great detail from the analysis of § 27. Moreover, further properties can be derived from the group situation; for instance we can now show that all minimal ideals are finite dimensional. Because of **27F** this will follow if we can show that every closed ideal contains a non-zero central element, and we give this proof now before restating the general results of § 27.

Lemma 1. *A continuous function $h(x)$ is in the center of the algebra $L^2(G)$ if and only if $h(xy) \equiv h(yx)$.*

Proof. $h \in$ center (L^2) if and only if

$$\int [h(xy)f(y^{-1}) - f(y)h(y^{-1}x)] \, dy = 0.$$

Replacing y by y^{-1} in the second term of the integrand and remembering that f is arbitrary, the result follows. This proof works just as well for the L^1 algebra of a unimodular locally compact group.

Lemma 2. *Every non-zero closed ideal $I \subset L^2$ contains a non-zero element of the center of L^2.*

Proof. (After Segal [44].) Choose any non-zero $g \in I$ and let $f = g * g^*$; f is continuous since $f(x) = (g_x, g)$ and since g_x is a continuous function of x as an element of L^2. Then $f(axa^{-1})$ is a continuous function of a if x is fixed, and, as an element of L^∞ is a continuous function of x (see **28B**). Thus $h(x) = \int f(axa^{-1}) \, da$ is continuous, and $h(e) = f(e) = \| g \|_2{}^2 > 0$, so that $h \neq 0$. Finally, if a is replaced by ay^{-1} in the integral for $h(yx)$, it follows from the right invariance of Haar measure that $h(yx) = h(xy)$, and hence that h is central. Moreover it is easy to see that $h \in I$, for $f \in I$ and $f_a{}^a \in I$ by **31F**, and it then follows from the Fubini theorem as in **31F** that $h \in I^{\perp\perp} = I$, q.e.d.

It now follows from **27F** that every minimal two-sided ideal N of $L^2(G)$ is finite-dimensional and contains an identity e. It then follows further that every function $f \in N$ is continuous, for $f = f * e$, or $f(x) = (f_x, e)$, and f_x as an element of L^2 is known to be a continuous function of x.

We now state the global structure theorem from § 27 as modified above.

Theorem. $L^2(G)$ *is the direct sum in the Hilbert space sense of its (mutually orthogonal) minimal two-sided ideals* N_α. *Each minimal two-sided ideal* N_α *is a finite-dimensional subspace of continuous functions. It has an identity* e_α, *and the projection on* N_α *of any function* $f \in L^2(G)$ *is* $f_\alpha = f * e_\alpha = e_\alpha * f$. *The regular maximal ideals of* L^2 *are the orthogonal complements of the minimal ideals, and every closed ideal in* L^2 *is at once the direct sum of the minimal ideals which it includes and the intersection of the maximal ideals which include it.*

The only central functions in N_α *are the scalar multiples of the identity* e_α, *and every central function* $f \in L^2$ *has a Fourier series expansion* $f = \sum \lambda_\alpha e_\alpha$, *where* $\lambda_\alpha e_\alpha = f * e_\alpha$. *If G is Abelian as well as compact, then each* N_α *is one-dimensional and the identities* e_α *are the characters of G.*

39B. The fine structure of a single minimal two-sided ideal N is also taken over from § 27, with additions. This time we state the theorem first.

Theorem. *If* $I_1, \cdots, I_n$ *are orthogonal minimal left ideals whose direct sum is a minimal two-sided ideal* N, *then* $I_1^*, \cdots, I_n^*$ *are minimal right ideals with the same property and* $I_i^* \cap I_j = I_i^* * I_j$ *is one-dimensional. Therefore there exist elements* $e_{ij} \in I_i^* \cap I_j$, *uniquely determined if* $i = j$ *and uniquely determined except for a scalar multiple of absolute value 1 in any case, such that* $e_{ij} * e_{kl} = e_{il}$ *if* $j = k$ *and* $= 0$ *if* $j \neq k$, $e_{ij} = e_{ji}^*$ *and* $e = \sum_1^n e_{ii}$. *The functions* $\{e_{ij}\}$ *form an orthogonal basis for* N *and the correspondence* $f \leftrightarrow (c_{ij})$, *where* $f = \sum c_{ij} e_{ij}$, *is an algebraic isomorphism between* N *and the full* $n \times n$ *matrix algebra, involution mapping into conjugate transposition. Furthermore* $\| e_{ij} \|_2^2 = n$ *and* $n e_{ik}(xy) = \sum_j e_{ij}(x) e_{jk}(y)$. *The left ideals* I_i *are isomorphic to each other (under the correspondence* $I_i e_{ij} = I_j$) *and to every other minimal left ideal in* N.

Proof. The only new facts concern the equation $n e_{ik}(xy) = \sum_j e_{ij}(x) e_{jk}(y)$. The left ideal I_k has the basis $e_{1k}, \cdots, e_{nk}$, and is invariant under left translation. Therefore, given x and i

there exist constants $c_{ij}(x)$ such that $e_{ik}(xy) = \sum_j c_{ij}(x)e_{jk}(y)$. Then

$$
\begin{aligned}
e_{ij}(x) &= [e_{i1} * e_{1j}](x) = \int e_{i1}(xy)e_{1j}(y^{-1})\, dy \\
&= \sum_m c_{im}(x) \int e_{m1}(y)\overline{e_{j1}(y)}\, dy = c_{ij}(x)\, \|\, e_{j1}\, \|_2{}^2 \\
&= c_{ij}(x)\, \|\, e_{11}\, \|_2{}^2,
\end{aligned}
$$

which gives the desired result if $\|\, e_{11}\, \|_2{}^2 = n$. But $\|\, e_{11}\, \|_2{}^4 = \|\, e_{11}\, \|_2{}^2 e_{11}(e) = \sum_{j=1}^n e_{1j}(x)e_{j1}(x^{-1}) = \sum_1^n |\, e_{1j}(x)\, |^2$, and integrating we get $\|\, e_{11}\, \|_2{}^4 = \sum_1^n \|\, e_{1j}\, \|_2{}^2 = n\|\, e_{11}\, \|_2{}^2$, or $\|\, e_{11}\, \|_2{}^2 = n$ as desired.

We have used the symbol e ambiguously for both the identity of the group G and of the ideal N.

39C. The identities e_α are called *characters* in the compact theory as well as in the Abelian theory; this again is terminology taken over from the classical theory of finite groups. It is not surprising that there is a functional equation for characters here as well as in the Abelian case.

Theorem. *If $f \in L^\infty(G)$, then there exists a scalar λ such that f/λ is a character if and only if there exists a scalar γ such that*

$$
\int f(xsx^{-1}t)\, dx = \frac{f(s)f(t)}{\gamma}.
$$

Proof. Given a character (or identity) e_α in a minimal ideal N_α, we have $\int e_\alpha(xsx^{-1}t)\, dx = \int e_\alpha(sx^{-1}tx)\, dx$ because e_α is central, and then we observe as in **39A**, Lemma 2, that because of the left invariance of Haar measure the integral is a central function of t for every fixed s. Since this central function belongs to N_α, it is a scalar multiple of e_α,

$$
\int e_\alpha(xsx^{-1}t)\, dx = k(s)e_\alpha(t).
$$

We evaluate $k(s)$ by taking $t = e$, giving $e_\alpha(s) = k(s)n$ and $\int e_\alpha(xsx^{-1}t)\, dx = e_\alpha(s)e_\alpha(t)/n$, where $n = e_\alpha(e) = \|\, e_\alpha\, \|_2{}^2$. If $f = \lambda e_\alpha$, then clearly $\int f(xsx^{-1}t)\, dx = f(s)f(t)/\gamma$, where $\gamma = n\lambda$.

Now suppose conversely that f satisfies this identity. As above $\int f(xsx^{-1}t)\,dx$ is a central function in s for every fixed t, and it follows from the assumed identity that f is central. If N_α is some minimal ideal on which f has a non-zero projection $f * e_\alpha = \lambda e_\alpha$, then

$$\iint f(xsx^{-1}t)e_\alpha(t^{-1})\,dx\,dt = \lambda \int e_\alpha(xsx^{-1})\,dx = \lambda e_\alpha(s)$$

$$\iint f(xsx^{-1}t)e_\alpha(t^{-1})\,dx\,dt = \frac{f(s)}{\gamma} \int f(t)e_\alpha(t^{-1})\,dt = f(s)\,\frac{(f,\,e_\alpha)}{\gamma}.$$

Since $(f,\,e_\alpha) = [f * e_\alpha](e) = \lambda e_\alpha(e) = \lambda n$, we see that $f(s) = \gamma e_\alpha(s)/n$, which is the desired result. Notice also that $\gamma = \lambda n$ as before.

39D. A function $f \in L^2$ is defined to be *almost invariant* if its mixed translates $f_a{}^b$ generate a finite-dimensional subspace of L^2. This subspace is then a two-sided ideal by **31F**, and is therefore a finite sum of minimal ideals. Thus f is almost invariant if and only if f lies in a finite sum of minimal two-sided ideals. The functions of a minimal ideal can be called minimal almost invariant functions, and the expansion $f = \sum f_\alpha$ can be viewed as the unique decomposition of f into minimal almost invariant functions.

The set of all almost invariant functions is a two-sided ideal, being, in fact, the *algebraic* sum $\sum_\alpha N_\alpha$ of all the minimal ideals. The ordinary pointwise product fg of two almost invariant functions f and g is also almost invariant. For if $\{f_i\}$ is a set of mixed translates of f spanning the invariant subspace generated by f and similarly for $\{g_j\}$, then the finite set of functions $\{f_i g_j\}$ spans the set of all translates of fg. Thus the set of almost invariant functions is a subalgebra of $\mathcal{C}(G)$.

Theorem. *Every continuous function on G can be uniformly approximated by almost invariant functions.*

Proof. If u is an approximate identity, then $u * f$ approximates f uniformly (for if $u = 0$ off V, then $|\,u * f(x) - f(x)\,| = |\int u(y)(f(y^{-1}x) - f(x))\,dy\,| \leqq \max_{y^{-1} \in V} \|\,f_y - f\,\|_\infty$, etc.), and it is sufficient to show that $u * f$ can be uniformly approximated by almost invariant functions. Let the sequence α_n include all the indices for which $f_\alpha \neq 0$ or $u_\alpha \neq 0$, and choose n so that $\| f - \sum_1^n f_{\alpha_i} \|_2 < \epsilon$ and $\| u - \sum_1^n u_{\alpha_i} \|_2 < \epsilon$. Then $\| u * f - \sum_1^n u_{\alpha_i} * f_{\alpha_i} \|_\infty = \| (u - \sum_1^n u_{\alpha_i}) * (f - \sum_1^n f_{\alpha_i}) \|_\infty < \epsilon^2$, q.e.d.

Another proof follows from the above remark that the vector space of almost invariant functions is closed under ordinary pointwise multiplication upon invoking the Stone-Weierstrass theorem.

39E. We conclude this section by showing that the ideal theory of $L^1(G)$ is identical with that of $L^2(G)$. Since $\| f \|_1 \leqq \| f \|_2$ the intersection with L^2 of any closed ideal I in L^1 is a closed ideal in L^2. If $I \neq 0$, then $I \cap L^2 \neq 0$. In fact, $I \cap L^2$ is dense (L^1) in I, for if $0 \neq f \in I$ and $u \in L^1 \cap L^2$ belongs to an approximate identity, then $u * f$ belongs to $I \cap L^2$ and approximates f in L^1. Since the L^1-closure of a closed ideal in L^2 is obviously a closed ideal in L^1, the mapping $I \to I \cap L^2$ is a one-to-one correspondence between the closed ideals of L^1 and L^2. The minimal ideals of L^2, being finite-dimensional, are identical with their L^1-closures and any L^1-closed ideal I is the direct sum in the L^1 sense of the minimal ideals it includes. In order to see this we approximate $f \in I$ by $g \in I \cap L^2$ and g by a finite sum of its components. This approximation improves in passing from the L^2 norm to the L^1 norm so that f is approximated L^1 by a finite sum of minimal almost invariant functions in I.

The other L^p algebras can be treated in a similar way, and a unified approach to the general kind of algebra of which these are examples has been given by Kaplansky [26], who calls such an algebra a *dual* algebra.

§ 40. REPRESENTATION THEORY

40A. Let T be any bounded representation of G. That is, T is a strongly continuous homomorphism $x \to T_x$ of G onto a group of uniformly bounded linear transformations $\{T_x\}$ of a normed linear space H into itself.

Theorem. *If H is a Hilbert space, with scalar product (u, v), then $[u, v] = \int (T_x u, T_x v)\, dx$ is an equivalent scalar product under which the transformations T_x are all unitary.*

Proof. It is clear that $[u, v]$ has all the algebraic properties of a scalar product except possibly that $u \neq 0$ implies $[u, u] \neq 0$. This follows from the fact that $(T_x u, T_x u)$ is a continuous function of x which is positive at $x = e$. Also $[T_y u, T_y v] = \int (T_x T_y u, T_x T_y v)\, dx = \int (T_{xy} u, T_{xy} v)\, dx = \int (T_x u, T_x v)\, dx = [u, v]$, so that the transformations T_x are all unitary in this scalar product. This completes the proof if H is finite-dimensional.

If B is a bound for the norms $\{\| T_x \| : x \in G\}$ and if $\| u \| = (u, u)^{1/2}$, $\||| u \|| = [u, u]^{1/2}$, we have at once that $\||| u \|| \leqq B \| u \|$. Since $u = T_{x^{-1}} v$ if $v = T_x u$, we have $\| u \| \leqq B \| T_x u \|$ and therefore $\| u \| \leqq B \||| u \||$. Therefore the norms $\| u \|$ and $\||| u \||$ are equivalent and the proof of the theorem is complete.

40B. We shall assume from now on that T is always a representation of G by unitary operators on a Hilbert space H. We know (see **32B, C**) that T is equivalent to a norm-decreasing, involution-preserving representation of L^1 by bounded operators on H. Since $L^2 \subset L^1$ and $\| f \|_2 \geqq \| f \|_1$ if G is compact and $\mu(G) = 1$, we thus have a norm-decreasing representation of $L^2(G)$, which we shall denote by the same letter T.

Our next arguments are valid for the *-representations of any H^*-algebra.

Theorem. *Every *-representation of an H^*-algebra A is uniquely expressible as a direct sum of (faithful) representations of certain of its minimal two-sided ideals.*

Proof. Let N_1 and N_2 be distinct (and hence orthogonal) minimal two-sided ideals of A and let H be the representation space. If $f \in N_1$, $g \in N_2$ and $u, v \in H$, then $(T_f u, T_g v) = (u, T_{f^* g} v) = 0$ since $f^* g = 0$. If we normalize H by throwing away the intersection of the nullspaces of all the operators T_f and for every index α set H_α equal to the closure of the union of the ranges of the operators T_f such that $f \in N_\alpha$, and set T^α

equal to the restriction of T to N_α, then it is clear that the sub-spaces H_α are orthogonal and add to H and that T is the direct sum of the representations T^α of the (simple) algebras N_α. Since each N_α is a minimal closed ideal, it follows that T^α is either 0 or faithful, and the kernel of T is exactly the sum of the ideals N_α for which $T^\alpha = 0$.

40C. The above representations T^α are not necessarily irreducible and we can now break them down further, although no longer in a unique manner. In our further analysis we shall focus attention on a single minimal ideal N and a faithful $*$-representation T of N by operators on a Hilbert space H, it being understood that the union of the ranges of the operators T_f is dense in H. Moreover, we shall assume that N is finite-dimensional; our proof that the irreducible constituents are all equivalent is not valid if N is infinite-dimensional.

Given a non-zero $v_1 \in H$, there exists a minimal left ideal $I \subset N$, with generating idempotent e, such that $T_e v_1 \neq 0$. Then $H_1 = \{T_f v_1 : f \in I\}$ is a finite-dimensional subspace of H, invariant under the operators T_f, $f \in N$, and operator isomorphic to I under the correspondence $f \to T_f v_1$. That the mapping is one-to-one follows in the usual way from the observation that, if it were not, then the kernel would be a non-zero proper sub-ideal of I. That it is an operator isomorphism follows from the equation $T_g(T_f v_1) = (T_g T_f)v_1 = T_{gf} v_1$, which establishes the desired relation between left multiplication by g on I and operating by T_g on H_1. Finally H_1 is irreducible, for if $v = T_f v_1 \in H_1$, then there exists $g \in N$ such that $gf = e$, so that $T_g v = T_e v_1$ and the subspace generated by v is the whole of H_1.

If H_1 is a proper subspace of H, we take $v_2 \in H_1^\perp$ and notice that then $(T_g v_2, T_f v_1) = (v_2, T_{g^* f}, v_1) = 0$, so that any minimal subspace derived from v_2 in the above manner is automatically orthogonal to H_1. Continuing in this way we break up H into a direct sum of finite-dimensional irreducible parts H_α. If T^α is the irreducible representation derived from T by confining it to H_α, then the preceding paragraph and **27E** imply that the T^α are all equivalent. We have proved: *Every faithful representation of a minimal closed ideal is a direct sum of equivalent irreducible representations.*

40D. We can now restrict ourselves to the case where T is an irreducible representation on a finite-dimensional Hilbert space V.

Let N be a minimal ideal on which T is an isomorphism and let (e_{ij}) be the matrix as described in **39B**. We shall use the notation $T(f)$ instead of T_f. $T(e_{kk})$ is a projection since $e_{kk}{}^2 = e_{kk}$; let V_k be its range. If v_1 is any non-zero vector of V_1, then $v_j = T(e_{j1})v_1$ is a vector in V_j, for $T(e_{jj})v_j = T(e_{jj}e_{j1})v_1 = T(e_{j1})v_1 = v_j$, and by a similar argument the vectors v_j have the property that $v_j = T(e_{ji})v_i$. Since $T(e_{ij})v_n = 0$ if $j \neq n$, the set v_j is linearly independent, and if V' is the subspace it spans, the equation $v_j = T(e_{ji})v_i$ shows that the coefficient matrix in the expansion of any $f \in N$ with respect to (e_{ij}) is identical with the matrix of $T(f)$ with respect to the basis $\{v_j\}$ in V'. Thus V' is invariant under T, and is operator isomorphic to any minimal left ideal in N. Also, since T is irreducible, $V' = V$.

If we now use the hypothesis that T is a *-representation, i.e., that $T(f^*) = T(f)^*$, then $\| v_j \|^2 = (T(e_{j1})v_1, T(e_{j1})v_1) = (v_1, T(e_{1j}e_{j1})v_1) = (v_1, T(e_{11})v_1) = (v_1, v_1) = \| v_1 \|^2$, and $(v_i, v_j) = (T(e_{i1})v_1, T(e_{j1})v_1) = (v_1, T(e_{1i}e_{j1})v_1 = 0$ if $i \neq j$. Therefore, the basis $\{v_i\}$ is orthonormal if $\| v_1 \| = 1$.

We now again assume that T arises from a representation T_x of a compact group. Since $U_x f = e^x * f$ on N, and $\overline{e^x(y)} = e(yx^{-1})$ $= \sum e_{ii}(yx^{-1}) = \sum_{i,j} e_{ij}(y)e_{ji}(x^{-1})/n = \sum [\overline{e_{ij}(x)}/n]e_{ij}(y)$, it follows that the matrix for T_x with respect to the orthonormal basis $\{v_i\}$ of V is $\overline{e_{ij}(x)}/n$. Any other orthonormal basis for V arises from the one considered here by a change of basis transformation, and under this transformation the functions e_{ij} give rise to a new set of similar functions related to the new basis. We have proved:

Theorem. *Let T be an irreducible representation of G by unitary transformations on the finite-dimensional Hilbert space V. Let T be extended to $L^2(G)$ as usual, and let N be the minimal ideal on which T is an isomorphism, the kernel of T being the regular maximal ideal $M = N^\perp$. If $c_{ij}(x)$ is the matrix of T_x with respect to any orthonormal basis for V, then the functions $e_{ij}(x) = nc_{ij}(x)$ are matrix generators for N as in **39B**, and every set of such matrix generators arises in this way.*

§ 41.   ALMOST PERIODIC FUNCTIONS

In this section we offer the reader an approach to almost periodic functions which as far as we know is new. We start off with the simple observation that under the uniform norm the left almost periodic functions on a topological group G form a commutative C^*-algebra (LAP) and therefore can be considered as the algebra $\mathcal{C}(\mathfrak{M})$ of all complex-valued continuous functions on a compact space $\mathfrak{M}$, the space of maximal ideals of LAP. The points of G define some of these ideals, a dense set in fact, and $\mathfrak{M}$ can be thought of as coming from G by first identifying points that are not separated by any $f \in$ LAP, then filling in new points corresponding to the remaining ideals, and finally weakening the topology to the weak topology defined by LAP. The critical lemma states that when G is considered as a subset of $\mathfrak{M}$ the operations of G are *uniformly continuous* in the topology of $\mathfrak{M}$ and hence uniquely extensible to the whole of $\mathfrak{M}$. We thus end up with a continuous homomorphism α of G onto a dense subgroup of a compact group $\mathfrak{M}$, with the property that the functions on G of the form $f(\alpha(s))$, $f \in \mathcal{C}(\mathfrak{M})$, are exactly the almost periodic functions on G (i.e., that the adjoint mapping α^* is an isometric isomorphism of $\mathcal{C}(\mathfrak{M})$ onto LAP).

41A. A left almost periodic function on a topological group G is a bounded continuous complex-valued function f whose set of left translates $S_f = \{f_s, s \in G\}$ is totally bounded under the uniform norm. The reader is reminded that a metric space S is totally bounded if and only if, for every $\epsilon > 0$, S can be covered by a finite number of spheres of radius ϵ. Also, a metric space is compact if and only if it is totally bounded and complete. Since the closure $\bar{S}_f$ of S_f in the uniform norm is automatically complete, we see that f is left almost periodic if and only if $\bar{S}_f$ is compact. Let LAP denote the set of all left almost periodic functions on G.

If g is a second function in LAP, the Cartesian product space $\bar{S}_f \times \bar{S}_g$ is compact and the mapping $\langle h, k \rangle \to h + k$ is continuous. Therefore the range of this mapping, the set of all sums of a function in S_f with a function in S_g, is compact, and the subset S_{f+g} is therefore totally bounded. We have proved that

LAP is closed under addition, and it is clear that the same proof works for multiplication.

Now suppose that $f^n \in$ LAP and f^n converges uniformly to f. Since $\| f^n_x - f_x \|_\infty = \| f^n - f \|_\infty$ the sequence f^n_x converges uniformly to f_x, uniformly over all $x \in G$. If, given ϵ, we choose f^n so that $\| f^n - f \|_\infty < \epsilon/3$ and choose the points x_i so that the set $\{ f^n_{x_i} \}$ is $\epsilon/3$-dense in S_{f^n}, then the usual combination of inequalities shows that $\{ f_{x_i} \}$ is ϵ-dense in S_f. That is, $f \in$ LAP, proving that LAP is complete.

If f is left almost periodic, then so are its real and imaginary parts and its complex conjugate. Thus LAP is closed under the involution $f \rightarrow \bar{f}$. We have proved:

Theorem. *Under the uniform norm the set of all left almost periodic functions on a topological group G is a commutative C^*-algebra with an identity.*

41B. It follows from the above theorem (and **26A**) that LAP is isomorphic and isometric to the space $\mathcal{C}(\mathfrak{M})$ of all complex-valued continuous functions on the compact Hausdorff space $\mathfrak{M}$ of its maximal ideals. Each point $s \in G$ defines a homomorphism $f \rightarrow f(s)$ of LAP onto the complex numbers and there is thus a natural mapping α of G into $\mathfrak{M}$ (where $\alpha(s)$ is the kernel of the homomorphism corresponding to s). The subsets of $\mathfrak{M}$ of the form $\{ M : |\hat{f}(M) - k | < \epsilon \}$ form a sub-basis for the topology of $\mathfrak{M}$. Since the inverse image under α of such a set is the open set $\{ s : | f(s) - k | < \epsilon \}$, the mapping α is continuous. $\alpha(G)$ is dense in $\mathfrak{M}$ since otherwise there exists a non-zero continuous function on $\mathfrak{M}$ whose correspondent on G is identically zero, a contradiction.

The mapping α is, of course, in general many-to-one and it is not *a priori* clear that the families of unseparated points in S are the cosets of a closed normal subgroup and that α therefore defines a group structure in the image $\alpha(G) \subset \mathfrak{M}$. Supposing this question to be satisfactorily settled there still remains the problem of extending this group structure to the whole of $\mathfrak{M}$. An analysis of these questions shows the need of a basic combinatorial lemma, which we present in the form of a direct elementary proof that a left almost periodic function is also right almost periodic.

The key device in this proof is the same as in the proof that the isometries of a compact metric space form a totally bounded set in the uniform norm.

Lemma. *If $f \in$ LAP and $\epsilon > 0$, there exists a finite set of points $\{b_i\}$ such that given any b one of the points b_i can be chosen so that $| f(xby) - f(xb_iy) | < \epsilon$ for every x and y.*

Proof. Let the finite set $\{f_{a_i}\}$ be $\epsilon/4$-dense among the left-translates of f. Given $b \in G$ and i, f_{a_ib} is within a distance $\epsilon/4$ of f_{a_j} for some j, and letting i vary we obtain an integer-valued function $j(i)$ such that $\| f_{a_ib} - f_{a_{j(i)}} \|_\infty < \epsilon/4$ for every i. For each such integer-valued function $j(i)$ let us choose one such b (if there is any) and let $\{b_j\}$ be the finite set so obtained. Then, by the very definition of $\{b_j\}$, for every b there exists one of the points b_k such that $\| f_{a_ib} - f_{a_ib_k} \|_\infty < \epsilon/2$ for all i. Since for any $x \in G$ we can find a_i so that $\| f_x - f_{a_i} \| < \epsilon/4$, we have $\| f_{xb} - f_{xb_k} \|_\infty \le \| f_{xb} - f_{a_ib} \|_\infty + \| f_{a_ib} - f_{a_ib_k} \|_\infty + | f_{a_ib_k} - f_{xb_k} \|_\infty < \epsilon/4 + \epsilon/2 + \epsilon/4 = \epsilon$. That is, for every $b \in G$ there exists one of the points b_k such that $| f(xby) - f(xb_ky) | < \epsilon$ for every x and y, q.e.d.

Corollary 1. *The set $\{f_{ij}\}$ defined by $f_{ij}(x) = f(b_ixb_j)$ is 2ϵ-dense in the set of all mixed translates $\{f_a{}^b\}$.*

Corollary 2. *If x and y are such that $| f(b_ixb_j) - f(b_iyb_j) | < \epsilon$ for all i and j, then $| f(uxv) - f(uyv) | < 5\epsilon$ for all u and $v \in G$.*

Proof. Given u and v we choose i and j so that $\| f(uxv) - f(b_ixb_j) \|_\infty < 2\epsilon$. This inequality combines with the assumed inequality to yield $| f(uxv) - f(uyv) | < 5\epsilon$, as desired.

41C. We now show that multiplication can be uniquely extended from G to the whole of $\mathfrak{M}$ by showing that, given M_1, $M_2 \in \mathfrak{M}$ and setting $\hat{x} = \alpha(x)$, if $\hat{x}_1 \to M_1$ and $\hat{x}_2 \to M_2$, then $\widehat{x_1x_2}$ converges. What we already know is that, given f and ϵ, there exist weak neighborhoods N_1 and N_2 of the points M_1 and M_2 respectively such that, if $\hat{x}_1, \hat{y}_1 \in N_1$ and $\hat{x}_2, \hat{y}_2 \in N_2$, then $| f(x_1x_2) - f(y_1y_2) | < \epsilon$; this follows from the second corollary of the paragraph above upon writing $f(x_1x_2) - f(y_1y_2) = f(x_1x_2) - f(x_1y_2) + f(x_1y_2) - f(x_2y_2)$. The family of sets $\{\widehat{x_1x_2} : x_1 \in N_1$

and $x_2 \in N_2$} obviously has the finite intersection property as N_1 and N_2 close down on M_1 and M_2 respectively, and, $\mathfrak{M}$ being compact, the intersection of their closures is non-empty. If M and M' are two of its points, then $|\hat{f}(M) - \hat{f}(M')| < \epsilon$ for every $\epsilon > 0$ and every $f \in \text{LAP}$ by the second sentence above, so that $M = M'$. We conclude that, given M_1 and M_2, there is a unique point M such that, for every weak neighborhood $N(M)$, there exist weak neighborhoods $N_1(M_1)$ and $N_2(M_2)$ such that $\hat{x}_1 \in N_1$ and $\hat{x}_2 \in N_2$ imply that $\widehat{x_1 x_2} \in N$. This shows first that the product $M_1 M_2$ is uniquely defined, and second, letting $\hat{x}_1$ and $\hat{x}_2$ converge to other points in N_1 and N_2, that the product $M_1 M_2$ is continuous in both factors. The reader who is familiar with the theory of uniform structures will realize that the above argument is a rather clumsy way of saying that $\widehat{x_1 x_2}$ is a uniformly continuous function of $\hat{x}_1$ and $\hat{x}_2$ in the uniform topology of $\mathfrak{M}$ and hence is uniquely extensible to the whole of $\mathfrak{M}$.

We still have to show the existence and continuity of inverses, i.e., loosely, that, if $\hat{x} \to M$, then x^{-1} converges. But by Corollary 2 of **41B**, given $f \in \text{LAP}$ and $\epsilon > 0$ there exists a weak neighborhood $N(M)$ such that, if $x, y \in N$, then $|f(uxv) - f(uyv)| < \epsilon$ for all u and v. Taking $u = x^{-1}$ and $v = y^{-1}$ this becomes $|f(y^{-1}) - f(x^{-1})| < \epsilon$ and proves exactly in the manner of the above paragraph the existence of a unique, continuous homeomorphism $M \to M^{-1}$ of M onto itself such that $\hat{x}^{-1} = (x^{-1})^{\wedge}$ for all $x \in G$. Thus $MM^{-1} = \lim (\hat{x}\hat{x}^{-1}) = \hat{e}$ and M^{-1} is the inverse of M. We have our desired theorem.

Theorem. *If G is any topological group, there is a compact group $\mathfrak{M}$ and a continuous homomorphism α of G onto a dense subgroup of $\mathfrak{M}$ such that a function f on G is left almost periodic if and only if there is a continuous function g on $\mathfrak{M}$ such that $f(s) \equiv g(\alpha(s))$ (i.e., $f = \alpha^* g$).*

There is a further remark which we ought to make in connection with the above theorem, namely, that the group $\mathfrak{M}$ is unique to within isomorphism. For if $\mathfrak{M}'$ is any second such group, with homomorphism β, then $\beta^{*-1}\alpha^*$ is an algebraic isomorphism of $\mathbb{C}(\mathfrak{M})$ onto $\mathbb{C}(\mathfrak{M}')$ and defines an associated homeomorphism γ of M onto M'. Since γ is obviously an extension of $\beta\alpha^{-1}$, which is an isomorphism on the dense subgroup $\alpha(G)$, γ itself is an iso-morphism. That is any two such groups M and M' are continu-ously isomorphic.

41D. Because of the above theorem the set of (left) almost periodic functions on a group G acquires all the algebraic struc-ture of the group algebra of a compact group. Thus any func-tion of LAP can be uniformly approximated by almost invariant functions, that is, by almost periodic functions whose translates generate a finite-dimensional subspace of L^∞. The more precise L^2 theorem according to which any almost periodic function has a Fourier series expansion into a series of uniquely determined *minimal* almost invariant functions is also available, but it has an apparent defect in that the Haar integral on the associated compact group seems to be necessary to define the scalar prod-uct. However, von Neumann [39] showed that an intrinsic defi-nition can be given for the mean $M(f)$ of an almost periodic function by proving the following theorem.

Theorem. *The uniformly closed convex set of functions generated by the left translates of an almost periodic function f contains ex-actly one constant function.*

The value of this constant function will, of course, be desig-nated $M(f)$. Von Neumann proved the existence and properties of $M(f)$ by completely elementary methods, using only the defi-nition of almost periodicity. It will be shorter for us in view of the above theory to demonstrate the existence of $M(f)$ by using the Haar integral on $\mathfrak{M}$. Since it will turn out that $M(f) = \int \hat{f}\, d\mu$, we will also have proved that the expansion theory based on the von Neumann mean is identical with that taken over from the associated compact group.

Proof of the theorem. The convex set generated by S_f is the set of finite sums $\sum c_i f(a_i x)$ such that $c_i > 0$ and $\sum c_i = 1$. We have to show that $M(f)$, and no other constant, can be ap-proximated uniformly by such sums. In view of the isometric imbedding of S_f into $\mathfrak{C}(\mathfrak{M})$ and the fact that $\alpha(G)$ is dense in $\mathfrak{M}$, it will be sufficient to prove the theorem on $\mathfrak{M}$. Given ϵ, choose V so that $|\,f(x) - f(y)\,| < \epsilon$ when $xy^{-1} \in V$ and choose $h \in L_V{}^+$.

Since $\mathfrak{M}$ is compact, there exists a finite set of points $a_1, \cdots, a_n$ such that $H(x) = \sum h(xa_i{}^{-1}) > 0$ on $\mathfrak{M}$. The functions $g_i = h^{a_i}/H$ thus have the property that $g_i = 0$ outside of Va_i and that

$$\sum g_i(x) \equiv 1. \quad \text{If } c_i = \int g_i, \text{ then}$$

$$\left| \int f(x)\,dx - \sum c_i f(a_i y) \right| = \left| \sum \int g_i(x)[f(xy) - f(a_i y)]\,dx \right| < \epsilon$$

since $| f(xy - f(a_i y) | < \epsilon$ when $g_i(x) \neq 0$. Thus $M(f) = \int f$ is approximated by the element $\sum c_i f_{a_i}$ of the convex set determined by S_f.

Conversely, if k is any number which can be so approximated, $\| k - \sum c_i f(a_i y) \|_\infty < \epsilon$, then $| k - \int f(y)\,dy | < \epsilon$ follows upon

integrating, so that $k = \int f = M(f)$, q.e.d.

Corollary. *Using $M(f)$ as an integral on the class of almost periodic functions, every almost periodic function has a unique L^2 expansion as a sum of minimal almost invariant functions.*

41E. If G is Abelian, then its associated compact group G_c is both compact and Abelian, and can be readily identified. The minimal almost invariant functions on G_c are, of course, its characters, and it follows that the minimal almost invariant functions on G are the characters of G, for $f(x) = h(\alpha(x))$ is clearly a character on G if h is a character on G_c. The adjoint of α thus determines an algebraic isomorphism between the character groups of G and G_c. Since $\hat{G}_c$ is discrete, it is completely identified as being the abstract group of characters on G with the discrete topology. Thus G_c is the character group of G^0, where G^0 is the character group $\hat{G}$, but in the discrete topology.

On an Abelian group the approximation and expansion theorem states that every almost periodic function can be uniformly approximated by linear combinations of characters and has a unique L^2 expansion as an infinite series of characters. In particular a function is almost periodic on the additive group of real numbers if and only if it can be uniformly approximated by linear combinations of exponentials, $\sum_1^m c_n e^{i\lambda_n x}$. This result will become the classical result when we have shown that the von Neumann definitions of almost periodicity and mean values are equivalent to the classical definitions. This we do in the next two theorems.

41F. Theorem. *A continuous function f on R is almost periodic if and only if, for every ϵ, there exists T such that in every interval of length T there is at least one point a such that $\| f - f_a \|_\infty < \epsilon$.*

Proof. We first rephrase the above condition as follows: for every ϵ there exists T such that for every a there exists $b \in [-T, T]$ such that $\| f_a - f_b \|_\infty < \epsilon$.

If f is assumed to be almost periodic and if the set $\{f_{a_i}\}$ is ϵ-dense in S_f, then we have only to take $[-T, T]$ as the smallest interval containing the points $\{a_i\}$ to meet the rephrased condition.

Conversely, suppose that T can always be found as in the rephrased condition. We first show that f is uniformly continuous. To do this we choose δ such that $| f(x + t) - f(x) | < \epsilon$ if $x \in [-T, T]$ and $| t | < \delta$. Then, given any x, we can choose $b \in [-T, T]$ so that $\| f_x - f_b \|_\infty < \epsilon$ and have

$$| f(x + t) - f(x) | \leqq | f(x + t) - f(b + t) |$$

$$+ | f(b + t) - f(b) | + | f(b) - f(x) | < 3\epsilon.$$

Thus $\| f - f_t \| < 3\epsilon$ if $| t | < \delta$.

Now let $\{a_1, \cdots, a_n\}$ be δ-dense on $[-T, T]$. Given a, we first choose $b \in [-T, T]$ so that $\| f_a - f_b \| < \epsilon$ and then a_i such that $| a_i - b | < \delta$, and therefore such that $\| f_b - f_{a_i} \|_\infty = \| f_{b-a_i} - f \|_\infty < 3\epsilon$ as above. Thus $\| f_a - f_{a_i} \| \leqq \| f_a - f_b \| + \| f_b - f_{a_i} \| < 4\epsilon$, proving that the set $\{f_{a_1}, \cdots, f_{a_n}\}$ is 4ϵ-dense in S_f and therefore that f is almost periodic.

41G. Theorem. $M(f) = \lim_{T \to \infty} \dfrac{1}{2T} \displaystyle\int_{-T}^{T} f(x)\, dx.$

Proof. Given ϵ, we choose the finite sets $\{c_i\}$ and $\{a_i\}$ such that $\| M(f) - \sum c_i f(x - a_i) \|_\infty < \epsilon$. It follows upon integrating that $| M(f) - (\sum c_i \int_{-T}^{T} f(x - a_i)\, dx)/2T | < \epsilon$. If we then choose $T > T_\epsilon = (\| f \|_\infty \cdot \max | a_i |)/\epsilon$, it follows that

$$\frac{1}{2T} \int_{-T}^{T} f(x - a_i)\, dx = \frac{1}{2T} \int_{-T-a_i}^{T-a_i} f(y)\, dy = \frac{1}{2T} \int_{-T}^{T} f(y)\, dy + \delta_i$$

where $| \delta_i | < | a_i | \| f \|/T < \epsilon$.   Therefore

$$| M(f) - \frac{1}{2T} \int_{-T}^{T} f(y)\, dy | < 2\epsilon$$

whenever $T > T_\epsilon$, q.e.d.

The alternate characterizations of almost periodicity and $M(f)$ given in the above two theorems are the classical ones for the real line. Thus the basic classical results now follow from our general theory.

41H. We ought not to close this section without some indication of the usual definition of the compact group G_c associated with G by its almost periodic functions. We have defined a left almost periodic function f to be such that the uniform closure $\bar{S}_f$ of the set S_f of its left translates is a compact metric space. Let $U_a{}^f$ be the restriction of the left translation operator U_a $(U_a f = f_{a^{-1}})$ to $\bar{S}_f$. $U_a{}^f$ is an isometry on $\bar{S}_f$ and an elementary (but not trivial) argument shows that the homomorphism $a \to U_a{}^f$ of G into the group of isometries on $\bar{S}_f$ is continuous, the uniform norm being used in the latter group. Now, it can be shown by an argument similar to that of **41B** that the isometries of a compact metric space form a compact group under the uniform norm. Thus the uniform closure G_f of the group $\{U_a{}^f : a \in G\}$ is a compact metric group, and the Cartesian product $\prod_f G_f$ is compact. Let $\bar{G}$ be the group of isometries T of LAP onto itself such that for every f the restriction T^f of T to S_f belongs to G_f. Each isometry in $\bar{G}$ determines a point of $\prod_f G_f$ and it is easy to see that the subset of $\prod_f G_f$ thus defined is closed and hence compact. That is, $\bar{G}$ itself is compact if it is given the following topology: a sub-basic neighborhood of an isometry $T_0 \in \bar{G}$ is specified in terms of a function $f \in$ LAP and a positive number ϵ as the set of isometries $T \in \bar{G}$ such that $\| T^f - T_0{}^f \| < \epsilon$. Also, under this topology the homomorphism α of G into $\bar{G}$ defined by $\alpha(s) = U_s$ is continuous.

If $f \in$ LAP, then $|f(x) - f(y)| \leq ||f_x - f_y||_\infty \leq || U_x{}^f - U_y{}^f ||$. That is, the function F defined by $F(U_x{}^f) = f(x)$ is uniformly continuous on a dense subset of G_f and hence is uniquely extensible to the whole of G_f. Since the projection of $\overline{G}$ onto $G_f (T \to T^f)$ is continuous, we may then consider F to be defined and continuous on the whole of $\overline{G}$. Moreover $F(\alpha(s)) = f(s)$ by definition. Thus for each $f \in$ LAP there exists $F \in \mathcal{C}(\overline{G})$ such that $f(s) = F(\alpha(s))$. Conversely any continuous F on the compact group $\overline{G}$ is almost periodic as we have observed earlier, and $F(\alpha(s))$ is therefore almost periodic on G. Thus $\overline{G}$ is (isomorphic to) the compact group associated with G. This method of constructing the compact group is essentially that given by Weil [48].

Chapter IX

SOME FURTHER DEVELOPMENTS

In this chapter we shall give a brief résumé of results in areas of this general field which are still only partially explored. Our aim is not so much to expound as to indicate a few of the directions in which the reader may wish to explore the literature further. It is suggested in this connection that the reader also consult the address of Mackey [35] in which many of these topics are discussed more thoroughly.

§ 42. NON-COMMUTATIVE THEORY

42A. The Plancherel theorem. We start with the problem of the generalization of the Plancherel theorem. It may have occurred to the reader that there is a great deal of similarity between the Fourier transform on an Abelian group and the expansion of a function into a sum of minimal almost invariant functions on a compact group. The following properties are shared by the two situations. There exists a locally compact topological space $\hat{G}$, and with each point $\alpha \in \hat{G}$ there is associated a minimal translation-invariant vector space V_α of bounded continuous functions on G. For each $f \in L^1$ there is uniquely determined a component function $f_\alpha \in V_\alpha$ such that, for each fixed $x, f_\alpha(x)$ is a continuous function of α vanishing at infinity. There is a measure μ on $\hat{G}$ such that, for each $f \in L^1 \cap \mathcal{P}(G), f_\alpha(x) \in L^1(\mu)$ for each x and $f(x) = \int f_\alpha(x) \, d\mu(\alpha)$. V_α is a (finite-dimensional) Hilbert space in a natural way, and, if $f \in L^1 \cap L^2(G)$, then $\| f \|_2{}^2 = \int \| f_\alpha \|^2 \, d\mu(\alpha)$ (Parseval's theorem). Thus the mapping $f \to f_\alpha$ can be extended to the whole of $L^2(G)$ preserving this equation, except that now f_α is determined modulo sets of μ-measure 0. This is the Plancherel theorem.

In each subspace V_α there is a uniquely determined function e_α characterized as being a central, positive definite, minimal almost invariant function. The component f_α in V_α of a function $f \in L^1$ is obtained as $f_\alpha = f * e_\alpha = U_f e_\alpha$, where U is the left regular representation. The subspaces V_α are invariant under U, and, if U^α is the restriction of U to V_α, then the function e_α is also given by $e_\alpha(x) = \text{trace } (U^\alpha{}_x)$. The component f_α has a similar alternate characterization, $f_\alpha(x) = \text{trace } (U^\alpha{}_x U^\alpha{}_f)$.

The Abelian case is simplified by the fact that the subspace V_α is one-dimensional, consisting of the scalar multiples of a single character $\alpha(x)$, but is the more complicated case in that a function f is expressed in terms of its component functions f_α by an *integration process* rather than a summation process. The compact group case is simpler in the latter respect; f is an ordinary sum of its components f_α. But the subspaces V_α are not one-dimensional.

It is, of course, impossible without some experience to select from the above rich collection of properties common to the compact groups and the Abelian locally compact groups those which could be expected to hold, and constitute a Plancherel theorem and its associated expansion theorem, on a locally compact group which is neither compact nor Abelian. However, the success of the above composite theory is in some sense due to the presence of sufficiently many central functions, and the extent to which it is a general phenomenon must be connected with the extent to which the general group has or does not have central functions, or at least central behavior of some kind.

What success has been achieved in the general situation has been based on von Neumann's reduction theory for rings of operators [40] which we now proceed to outline.

We suppose given a measure space S, with each point $\alpha \in S$ an associated Hilbert space H_α, and a vector space of functions f on S such that $f(\alpha) \in H_\alpha$ for every $\alpha \in S$ and $\| f(\alpha) \| \in L^2(S)$.

The ordinary L^2-completion of this vector space is then a Hilbert space H of functions f of the above type, and we write $H = \int H_\alpha \, d\mu(\alpha)$, where μ is the given measure on S.

If we are given a separable Hilbert space H and a complete Boolean algebra $\mathcal{B}$ of commuting projections on H, then the standard theory of unitary equivalence can be interpreted as giving a direct integral representation of H, $H = \int H_\alpha \, d\mu(\alpha)$, such that the projections of $\mathcal{B}$ are exactly those projections E_C associated with the measurable subsets C of S, $E_C f = \int_C f(\alpha) \, d\mu$.

If $H = \int H_\alpha \, d\mu(\alpha)$ and if T is a bounded operator on H, then T is defined to be a direct integral if and only if there exists for each α a bounded operator T_α on H_α such that, if $x = \int x_\alpha \, d\mu(\alpha) \in H$, then $T(x) = \int (T_\alpha x_\alpha) \, d\mu(\alpha)$. One of von Neumann's basic theorems is that, if T commutes with all the projections E_C associated with the measurable subsets of the index space S, then T is a direct integral.

An algebra of bounded operators on a Hilbert space H is said to be weakly closed if it is a closed subset of the algebra of all bounded operators in the weak topology generated by the functions $f_{x,y}(T) = (Tx, y)$ and if it is closed under the involution $T \to T^*$. For any set $\mathcal{A}$ of operators, $\mathcal{A}'$ is defined to be the set of all operators commuting with every operator in $\mathcal{A}$. Then it is a fundamental fact that a $*$-closed set $\mathcal{A}$ of operators is a weakly-closed algebra if and only if $\mathcal{A} = \mathcal{A}''$.

If $\mathcal{A}$ is a weakly-closed algebra, then $\mathcal{A} \cap \mathcal{A}'$ is clearly the *center* of $\mathcal{A}$. The center of $\mathcal{A}$ consists only of scalar multiples of the identity if and only if the union $\mathcal{A} \cup \mathcal{A}'$ generates the algebra $B(H)$ of all bounded operators on H; in this case $\mathcal{A}$ is said to be a *factor*. Von Neumann and Murray's dimensional analysis of factors, with the resulting classification of all factors into types I_n, I_∞, II_1, II_∞ and III, is one of the basic results of the theory, but will not be gone into here.

Now let α be any weakly-closed algebra of bounded operators on H and let $H = \int H_\alpha \, d\mu(\alpha)$ be the expression of H as a direct integral with respect to the center of α. That is, the projections E_C associated with the measurable subsets C of the index space are exactly the projections in the center of α. Let T_n be a sequence of operators in α which generates α, let $T_n = \int T_{n,\alpha} \, d\mu(\alpha)$ be the direct integral expression of T_n and let α_α be the weakly-closed algebra of bounded operators on H_α generated by the sequence $T_{n,\alpha}$. Then α is the direct integral of the algebras α_α in the sense that a bounded operator T on H belongs to α if and only if it is a direct integral $\int T_\alpha \, d\mu(\alpha)$ where $T_\alpha \in \alpha_\alpha$.

Another of von Neumann's fundamental theorems is that in this direct integral decomposition of α almost all the algebras α_α are factors, and that α' is the direct integral of the commuting factors α_α'. Godement, Mautner and Segal ([21], [37], [38], [45]; Godement's major work in this direction is unpublished at the time of writing) have considered $L^1(G)$ on a separable unimodular locally compact group G as both a right and left operator algebra on the Hilbert space $L^2(G)$ and have observed the fundamental fact that, if $\mathcal{L}$ and $\mathcal{R}$ are the weak closures of these operator algebras, then $\mathcal{L}' = \mathcal{R}$ and $\mathcal{R}' = \mathcal{L}$. Thus the intersection $\mathcal{L} \cap \mathcal{R}$ is the center of each algebra, and von Neumann's reduction theory implies that $H = L^2(G)$ can be expressed as the direct integral of Hilbert spaces H_α in such a way that each operator $A \in \mathcal{L}$ is the direct integral of operators A_α acting on H_α, and the weakly-closed algebra $\mathcal{L}_\alpha$ generated over H_α by the operators A_α is, for almost all α, a factor.

This analysis of $\mathcal{L}$ into factors is what corresponds to the analysis of $L^1(G)$ into minimal parts in the simpler theories, and a Parseval formula is valid; if $f \in L^1 \cap L^2$, then $\| f \|_2{}^2 = \int \mathrm{trace}\ U_f{}^\alpha U_{f^*}{}^\alpha \, d\alpha$. However, these oversimplified statements conceal a multitude of complications and the reader is referred to the literature cited above for details. Godement has so framed his work that it applies to other two-sided representations in addition to the two-sided regular representation mentioned above, and he has introduced a topology into his index space, starting with the observation that the common center $\mathcal{L} \cap \mathcal{R}$ is a commutative C^*-Banach algebra and therefore is isometric to the space $\mathcal{C}(\mathfrak{M})$ of all continuous complex-valued functions on its compact Hausdorff space of maximal ideals. The points $M \in \mathfrak{M}$, subject to certain later identifications, are Godement's indices α. Kaplansky [28] has investigated a much narrower class of groups but obtains a much more satisfactory theory as a result.

42B. Representation theory. The above Plancherel theorem can be interpreted as the analysis of the left regular representation of G into factor representations, and suggests the corresponding problem for arbitrary representations. This is not the same thing as analyzing a representation into irreducible parts, but seems to be in many ways the more natural objective. Very little is known as yet about the general representation theory of locally compact groups beyond the basic facts that sufficiently many irreducible representations always exist and are in general infinite-dimensional.

However, there is considerable literature on the representation theory of special groups and of general groups with respect to special situations. The Russians have published a series of papers (of which we have listed only one, [14]) analyzing the irreducible representations of certain finite-dimensional Lie groups. Mackey [36] has been able to subsume many of these special results under a generalization of the theory of induced representations from its classical setting in finite groups to separable unimodular locally compact groups. The representation theory of Lie groups has also been explored in its connection with the representation theory of Lie algebras.

We saw in § 32 how representations of groups are intimately associated with representations of the corresponding L^1-algebras. In fact, group algebras were first introduced as an aid in the study of the representations of finite groups (see [32] for a short account). Of course, the representation theory of general Banach algebras is a natural subject for study, quite apart from such motivations. Unfortunately, the results are meagre. We have cited two. One is the theorem of **26F** on the representations of a commutative self-adjoint Banach algebra, and the other is the theorem of Gelfand and Neumark [13] to the effect that every C^*-algebra is isometric and isomorphic to an algebra of bounded operators on a Hilbert space. These are, of course, very restricted conclusions, but little is known of general representation theory, even for such a special case as that of a Banach algebra with an involution (see [28]).

§ 43. COMMUTATIVE THEORY

43A. The Laplace transform. Mackey [34] has observed that the kernel $e^{(x+iy)s}$ of the classical Laplace transform can be given a characterization valid on a general locally compact Abelian group in much the same way as Fourier transform theory is shifted to this setting by replacing the kernel e^{iys} by the general character function $\alpha(s)$. The following description is an over-simplification but will illustrate the kind of step to be taken. We replace the mapping $s \rightarrow (x + iy)s$, where x and y are fixed, by a continuous homomorphism v of an Abelian topological group G into the additive group of the complex numbers. The set of all such homomorphisms is clearly a complex vector space V and an element $v \in V$ will be called simply a vector. The kernel $e^{(x+iy)s}$ is then replaced by the kernel $e^{v(s)}$, $v \in V$, $s \in G$, and the generalized Laplace transform is the mapping $f \rightarrow F$ where

$$F(v) = \int e^{v(s)} f(s) \, ds.$$

A complex-valued function defined on a suitable subset of V will be said to be analytic at v if for every $u \in V$ the limit $F(v, u)$ of of $F(v + \lambda u) - F(v)/\lambda$ exists as the complex variable λ approaches 0. Leaving aside questions of existence, we see that formally the Laplace transform F of a function f on G is analytic and

$$F(v, u) = \int e^{v(s)} u(s) f(s) \, ds.$$

Mackey has shown that in a suitably restricted L^2 sense a function F defined on V is analytic if and only if it is a Laplace transform. Many other theorems of Laplace transform theory have

natural extensions, and the abstract setting of the theory seems
to lead to a fruitful point of view. In particular, it is expected
that some of the theory of functions of several complex variables
will be illuminated by this general formulation.

43B. Beurling's algebras. Another problem which leads to this
same generalization of the Laplace transform, and which has
been discussed in a special case by Beurling [4] is the theory of
subalgebras of $L^1(G)$ which are themselves L^1 spaces with respect
to measures "larger than" Haar measure. The example considered in the second part of **23D** is a case in point. More generally,
suppose that ω is a positive weight function on an arbitrary
locally compact Abelian group such that $\omega(xy) \leqq \omega(x)\omega(y)$ for
all $x, y \in G$. Then the set L^ω of measurable functions f such that
$\int | f |\omega < \infty$ is the space $L^1(\mu)$ where $\mu(A) = \int_A \omega(s) \, ds$, and the
inequality on ω assumed above implies that L^ω is closed under
convolution and is commutative Banach algebra. A simple condition guaranteeing that L^ω is a subset of $L^1(G)$ is that ω be
bounded below by 1. Then every regular maximal ideal of $L^1(G)$
defines one for L^ω (see **24B**) and the maximal ideal space of $L^1(G)$
is identifiable with a (closed) subset of that of L^ω, generally a
proper subset. The homomorphisms of L^ω onto the complex num-
bers are given by functions $\theta_M \in L^\infty(G), \hat{f}(M) = \int f(s)\overline{\theta_M(s)}\omega(s) \, ds$,
and a simple argument as in **23D** shows that the product $\theta_M(s)\omega(s)$
is a *generalized character*, that is, a continuous homomorphism of
G into the multiplicative group of the non-zero complex numbers, and therefore that its logarithm, if it can be taken, is what
we have called in the above paragraph a vector. The transforms
$\hat{f}$ of the elements $f \in L^\omega$ are thus generally to be considered as
bilateral Laplace transforms, and the theory of such algebras L^ω
merges with the general Laplace transform theory.

43C. Tauberian theorems. In case ω is such that the convolu-
tion algebra $L^\omega = \{f : \int | f |\omega < \infty\}$ has the *same* maximal ideal
space as the larger algebra $L^1(G)$ it is natural to consider whether
ideal theoretic theorems such as Wiener's Tauberian theorem remain valid. Questions of this nature were taken up by Beurling in

1938 [4] for the special case of the real line. Beurling's basic theorem in this direction is roughly that, if

$$\int_{-\infty}^{\infty} \frac{|\log \omega(x)|}{1 + x^2} \, dx < \infty,$$

then L^ω is regular and the Tauberian theorem holds: every closed proper ideal lies in a regular maximal ideal.

It should be remarked that Beurling proved and leaned heavily on the formula: $\| \hat{f} \|_\infty = \lim \| f^n \|^{1/n}$. His proof uses the properties of the real numbers in an essential way and does not generalize; Gelfand's later proof of the same formula for a general Banach algebra is based on an entirely different approach.

Returning to our main theme, we notice that the Tauberian theorem on a locally compact Abelian group can be rephrased as the assertion that any proper closed invariant subspace of $L^1(G)$ is included in a maximal one, and then can be rephrased again, via orthogonality, as the statement that every non-void translation-invariant, weakly-closed subspace of L^∞ includes a minimal one (the scalar multiples of a character). In another paper [5] Beurling has demonstrated this form of the Tauberian theorem for a different space, the space of bounded continuous functions on the real line, under the topology characterized by uniform convergence on compact sets together with the convergence of the uniform norm itself. In addition he showed that unless the closed invariant subspace is one-dimensional it contains *more* than one exponential. This is clearly the analogue of the Segal-Kaplansky-Helson theorem, and suggests that we leave the general Tauberian problem in the following form. It is required to find conditions which will imply that a closed ideal in a commutative Banach algebra A shall be the kernel of its hull, or, dually, that a weakly-closed invariant subspace of A^* shall be spanned by the homomorphisms which it contains. The condition can be either on the algebra or on the subspace, and in the case of groups it can be on the measure defining the convolution algebra or on the topology used on the conjugate space.

43D. Primary ideals. Although a closed ideal in a commutative Banach algebra is not, in general, the intersection of maximal ideals (i.e., not the kernel of its hull) there are two spectacular cases in which it is known that every closed ideal is the intersection of *primary* ideals, a primary ideal being defined as an ideal which is included in only one maximal ideal. The first (due to Whitney [49]) is the algebra of functions of class C^r on a region of Euclidean n-space, its topology taking into account all derivatives up to and including order r.

The second (due to Schwartz [43]) is the algebra of all bounded complex-valued Baire measures on the real line which are supported by compact sets, with the topology which it acquires as the conjugate space of the space of all (not necessarily bounded) continuous functions under the topology defined by uniform convergence on compact sets. The continuous homomorphisms of this algebra into the complex numbers are given by exponentials $e^{\alpha t}$ $(\hat{\mu}(\alpha) = \int e^{\alpha t}\, d\mu(t))$ and the regular maximal ideal space is thus identified with the complex plane, the transforms $\hat{\mu}$ forming an algebra of entire functions. Every primary ideal of this algebra is specified by a complex number α_0 and an integer n, and consists of the measures μ such that $\hat{\mu}$ and its first n derivatives vanish at α_0. And every closed ideal is an intersection of primary ideals.

In both of these examples the presence and importance of primary ideals are connected with differentiation.

43E. We conclude with a brief mention of a remarkable analogy which has been studied by Levitan, partly in collaboration with Powsner, in a series of papers of which we have listed only [31]. Levitan observed that the eigen-function expansions of Sturm-Liouville differential equation theory can be regarded as a generalization of the Fourier transform theory of group algebras as discussed in Chapter VII. Specifically, let us consider the differential equation $y'' - (\rho(x) - \lambda)y = 0$ on the positive real axis, with a boundary condition at $x = 0$. With this ordinary differential equation we associate the partial differential equation $u_{xx} - u_{yy} - (\rho(x) - \rho(y))u = 0$ with initial conditions $u(x, 0) = f(x)$, $u_y(x, 0) = 0$, where ρ and f are even functions. The solution $u(x, y) = [T_y f](x)$ is regarded as a generalization of the translation of $f(x)$ through y. If generalized convolution is defined in the obvious way an L^1 convolution algebra is obtained which has very far-reaching analogies with the L^1 group algebras. The following topics can be developed: generalized characters (eigen-functions), positive definite functions, Bochner's theorem, the L^1-inversion theorem and the unique measure on the transform space, and the Plancherel theorem. The theory is valid not only when the spectrum is discrete but also in the limit point case when the spectrum is continuous, although it is not yet fully worked out in the latter case. The reader is referred to the literature for further details.

BIBLIOGRAPHY

The following bibliography contains only those items explicitly referred to in the text. For a more complete bibliography see the book of Weil ([48] below) for the years preceding 1938, and the address of Mackey ([35] below) for subsequent years.

1. W. AMBROSE, *Structure theorems for a special class of Banach algebras*, Trans. Amer. Math. Soc. **57**, 364–386 (1945).

2. R. ARENS, *On a theorem of Gelfand and Neumark*, Proc. Nat. Acad. Sci. U.S.A. **32**, 237–239 (1946).

3. S. BANACH, *Théorie des opérations linéaires*, Warsaw, 1932.

4. A. BEURLING, *Sur les intégrales de Fourier absolument convergentes et leur application à une transformation fonctionnelle*, Congrès des Mathématiques à Helsingfors, 1938.

5. A. BEURLING, *Un théorème sur les fonctions bornées et uniformement continues sur l'axe réel*, Acta Math. **77**, 127–136 (1945).

6. H. BOHNENBLUST and A. SOBCZYK, *Extensions of functionals on complex linear spaces*, Bull. Amer. Math. Soc. **44**, 91–93 (1938).

7. H. CARTAN, *Sur la mesure de Haar*, C. R. Acad. Sci. Paris **211**, 759–762 (1940).

8. H. CARTAN and R. GODEMENT, *Théorie de la dualité et analyse harmonique dans les groupes abeliens localement compacts*, Ann. Sci. Ecole Norm, Sup. (3) **64**, 79–99 (1947).

9. P. DANIELL, *A general form of integral*, Ann. of Math. **19**, 279–294 (1917).

10. V. DITKIN, *On the structure of ideals in certain normed rings*, Ucenye Zapiski Moskov. Gos. Univ. Mat. **30**, 83–130 (1939).

11. J. DIXMIER, *Sur un théorème de Banach*, Duke Math. J. **15**, 1057–1071 (1948).

12. I. GELFAND, *Normierte Ringe*, Rec. Math. (Mat. Sbornik) N.S. **9**, 3–24 (1941).

13. I. GELFAND and M. NEUMARK, *On the imbedding of normed rings into the ring of operators in Hilbert space*, Rec. Math. N.S. **12** (54), 197–213 (1943).

14. I. GELFAND and M. NEUMARK, *Unitary representations of the Lorentz group*, Izvestiya Akademii Nauk SSSR. Ser. Mat. **11**, 411–504 (1947).

15. I. GELFAND and D. RAIKOV, *On the theory of character of commutative topological groups*, C. R. (Doklady) Acad. Sci. URSS **28**, 195–198 (1940).

16. I. GELFAND and D. RAIKOV, *Irreducible unitary representations of locally bicompact groups*, Rec. Math. (Mat. Sbornik) N.S. **13**, 301–316 (1943).

17. I. GELFAND, D. RAIKOV, and G. SILOV, *Commutative normed rings*, Uspehi Matematiceskih Nauk. N.S. **2**, 48–146 (1946).

18. A. GLEASON, *A note on locally compact groups*, Bull. Amer. Math. Soc. **55**, 744–745 (1949).

19. R. GODEMENT, *Les fonctions de type positif et la théorie des groupes*, Trans. Amer. Math. Soc. **63**, 1–84 (1948).

20. R. GODEMENT, *Sur la théorie des représentations unitaires*, Ann. of Math. (2) **53**, 68–124 (1951).

21. R. GODEMENT, *Memoire sur la théorie des caractères dans les groupes localement compacts unimodulaires*, Journal de Math. Purée e Appliqué **30**, 1–110 (1951).

22. A. HAAR, *Der Massbegriff in der Theorie der kontinuierlichen Gruppen*, Ann. of Math. **34**, 147–169 (1933).

23. P. HALMOS, *Measure Theory*, Van Nostrand, New York, 1950.

24. H. HELSON, *Spectral synthesis of bounded functions*, Ark. Mat. **1**, 497–502 (1951).

25. N. JACOBSON, *The radical and semi-simplicity for arbitrary rings*, Amer. J. Math. **67**, 300–320 (1945).

26. I. KAPLANSKY, *Dual rings*, Ann. of Math. (2) **49**, 689–701 (1948).

27. I. KAPLANSKY, *Primary ideals in group algebras*, Proc. Nat. Acad. Sci. U.S.A. **35**, 133–136 (1949).

28. I. KAPLANSKY, *The structure of certain operator algebras*, Trans. Amer. Math. Soc. **70**, 219–255 (1951).

29. M. KREIN, *Sur une généralisation du théorème de Plancherel au cas des intégrales de Fourier sur les groupes topologiques commutatifs*, C. R. (Doklady) Acad. Sci. URSS (N.S.) **30**, 484–488 (1941).

30. M. KREIN and D. MILMAN, *On extreme points of regular convex sets*, Studia Mathematica **9**, 133–138 (1940).

31. B. M. LEVITAN, *The application of generalized displacement operators to linear differential equations of the second order*, Uspehi Matem. Nauk (N.S.) **4**, no. 1 (29), 3–112 (1949).

32. D. E. LITTLEWOOD, *The Theory of Group Characters and Matrix Representations of Groups*, Oxford Univ. Press, New York, 1940.

33. L. H. LOOMIS, *Haar measure in uniform structures*, Duke Math. J. **16**, 193–208 (1949).

34. G. MACKEY, *The Laplace transform for locally compact Abelian groups*, Proc. Nat. Acad. Sci. U.S.A. **34**, 156–162 (1948).

35. G. MACKEY, *Functions on locally compact groups*, Bull. Amer. Math. Soc. **56**, 385–412 (1950).

36. G. MACKEY, *On induced representations of groups*, Amer. J. Math. **73**, 576–592 (1951).

37. F. I. MAUTNER, *Unitary representations of locally compact groups I*, Ann. of Math. **51**, 1–25 (1950).

38. F. I. MAUTNER, *Unitary representations of locally compact groups II*, Ann. of Math., to be published.

39. J. VON NEUMANN, *Almost periodic functions in a group*, Trans. Amer. Math. Soc. **36**, 445–492 (1934).

40. J. VON NEUMANN, *On rings of operators. Reduction theory*. Ann. of Math. **50**, 401–485 (1949).

41. D. A. RAIKOV, *Generalized duality theorem for commutative groups with an invariant measure*, C. R. (Doklady) Acad. Sci. URSS (N.S.) **30**, 589–591 (1941).

42. D. A. Raikov, *Harmonic analysis on commutative groups with the Haar measure and the theory of characters*, Trav. Inst. Math. Stekloff 14 (1945).

43. L. Schwartz, *Théorie générale des fonctions moyennepériodiques*, Ann. of Math. **48**, 857–929 (1947).

44. I. E. Segal, *The group algebra of a locally compact group*, Trans. Amer. Math. Soc. **61**, 69–105 (1947).

45. I. E. Segal, *An extension of Plancherel's formula to separable unimodular groups*, Ann. of Math. (2) **52**, 272–292 (1950).

46. M. H. Stone, *The generalized Weierstrass approximation theorem*, Math. Mag. **21**, 167–184, 237–254 (1948).

47. M. H. Stone, *Notes on integration: II*, Proc. Nat. Acad. Sci. U.S.A. **34**, 447–455 (1948).

48. A. Weil, *L'intégration dans les groupes topologiques et ses applications*, Paris, 1938.

49. H. Whitney, *On ideals of differentiable functions*, Amer. J. Math. **70**, 635–658 (1948).

INDEX

The references in the index are, principally, either to the definitions of terms or to the statements of theorems.

Mathematics

FUNCTIONAL ANALYSIS (Second Corrected Edition), George Bachman and Lawrence Narici. Excellent treatment of subject geared toward students with background in linear algebra, advanced calculus, physics and engineering. Text covers introduction to inner-product spaces, normed, metric spaces, and topological spaces; complete orthonormal sets, the Hahn-Banach Theorem and its consequences, and many other related subjects. 1966 ed. 544pp. 6⅛ x 9¼. 0-486-40251-7

DIFFERENTIAL MANIFOLDS, Antoni A. Kosinski. Introductory text for advanced undergraduates and graduate students presents systematic study of the topological structure of smooth manifolds, starting with elements of theory and concluding with method of surgery. 1993 edition. 288pp. 5⅜ x 8½. 0-486-46244-7

VECTOR AND TENSOR ANALYSIS WITH APPLICATIONS, A. I. Borisenko and I. E. Tarapov. Concise introduction. Worked-out problems, solutions, exercises. 257pp. 5⅝ x 8¼. 0-486-63833-2

AN INTRODUCTION TO ORDINARY DIFFERENTIAL EQUATIONS, Earl A. Coddington. A thorough and systematic first course in elementary differential equations for undergraduates in mathematics and science, with many exercises and problems (with answers). Index. 304pp. 5⅜ x 8½. 0-486-65942-9

FOURIER SERIES AND ORTHOGONAL FUNCTIONS, Harry F. Davis. An incisive text combining theory and practical example to introduce Fourier series, orthogonal functions and applications of the Fourier method to boundary-value problems. 570 exercises. Answers and notes. 416pp. 5⅜ x 8½. 0-486-65973-9

COMPUTABILITY AND UNSOLVABILITY, Martin Davis. Classic graduate-level introduction to theory of computability, usually referred to as theory of recurrent functions. New preface and appendix. 288pp. 5⅜ x 8½. 0-486-61471-9

AN INTRODUCTION TO MATHEMATICAL ANALYSIS, Robert A. Rankin. Dealing chiefly with functions of a single real variable, this text by a distinguished educator introduces limits, continuity, differentiability, integration, convergence of infinite series, double series, and infinite products. 1963 edition. 624pp. 5⅜ x 8½. 0-486-46251-X

METHODS OF NUMERICAL INTEGRATION (SECOND EDITION), Philip J. Davis and Philip Rabinowitz. Requiring only a background in calculus, this text covers approximate integration over finite and infinite intervals, error analysis, approximate integration in two or more dimensions, and automatic integration. 1984 edition. 624pp. 5⅜ x 8½. 0-486-45339-1

INTRODUCTION TO LINEAR ALGEBRA AND DIFFERENTIAL EQUATIONS, John W. Dettman. Excellent text covers complex numbers, determinants, orthonormal bases, Laplace transforms, much more. Exercises with solutions. Undergraduate level. 416pp. 5⅜ x 8½. 0-486-65191-6

RIEMANN'S ZETA FUNCTION, H. M. Edwards. Superb, high-level study of landmark 1859 publication entitled "On the Number of Primes Less Than a Given Magnitude" traces developments in mathematical theory that it inspired. xiv+315pp. 5⅜ x 8½. 0-486-41740-9

CALCULUS OF VARIATIONS WITH APPLICATIONS, George M. Ewing. Applications-oriented introduction to variational theory develops insight and promotes understanding of specialized books, research papers. Suitable for advanced undergraduate/graduate students as primary, supplementary text. 352pp. 5³/₈ x 8¹/₂.
0-486-64856-7

MATHEMATICIAN'S DELIGHT, W. W. Sawyer. "Recommended with confidence" by *The Times Literary Supplement*, this lively survey was written by a renowned teacher. It starts with arithmetic and algebra, gradually proceeding to trigonometry and calculus. 1943 edition. 240pp. 5³/₈ x 8¹/₂.
0-486-46240-4

ADVANCED EUCLIDEAN GEOMETRY, Roger A. Johnson. This classic text explores the geometry of the triangle and the circle, concentrating on extensions of Euclidean theory, and examining in detail many relatively recent theorems. 1929 edition. 336pp. 5³/₈ x 8¹/₂.
0-486-46237-4

COUNTEREXAMPLES IN ANALYSIS, Bernard R. Gelbaum and John M. H. Olmsted. These counterexamples deal mostly with the part of analysis known as "real variables." The first half covers the real number system, and the second half encompasses higher dimensions. 1962 edition. xxiv+198pp. 5³/₈ x 8¹/₂.
0-486-42875-3

CATASTROPHE THEORY FOR SCIENTISTS AND ENGINEERS, Robert Gilmore. Advanced-level treatment describes mathematics of theory grounded in the work of Poincaré, R. Thom, other mathematicians. Also important applications to problems in mathematics, physics, chemistry and engineering. 1981 edition. References. 28 tables. 397 black-and-white illustrations. xvii + 666pp. 6¹/₈ x 9¹/₄.
0-486-67539-4

COMPLEX VARIABLES: Second Edition, Robert B. Ash and W. P. Novinger. Suitable for advanced undergraduates and graduate students, this newly revised treatment covers Cauchy theorem and its applications, analytic functions, and the prime number theorem. Numerous problems and solutions. 2004 edition. 224pp. 6¹/₂ x 9¹/₄.
0-486-46250-1

NUMERICAL METHODS FOR SCIENTISTS AND ENGINEERS, Richard Hamming. Classic text stresses frequency approach in coverage of algorithms, polynomial approximation, Fourier approximation, exponential approximation, other topics. Revised and enlarged 2nd edition. 721pp. 5³/₈ x 8¹/₂.
0-486-65241-6

INTRODUCTION TO NUMERICAL ANALYSIS (2nd Edition), F. B. Hildebrand. Classic, fundamental treatment covers computation, approximation, interpolation, numerical differentiation and integration, other topics. 150 new problems. 669pp. 5³/₈ x 8¹/₂.
0-486-65363-3

MARKOV PROCESSES AND POTENTIAL THEORY, Robert M. Blumental and Ronald K. Getoor. This graduate-level text explores the relationship between Markov processes and potential theory in terms of excessive functions, multiplicative functionals and subprocesses, additive functionals and their potentials, and dual processes. 1968 edition. 320pp. 5³/₈ x 8¹/₂.
0-486-46263-3

ABSTRACT SETS AND FINITE ORDINALS: An Introduction to the Study of Set Theory, G. B. Keene. This text unites logical and philosophical aspects of set theory in a manner intelligible to mathematicians without training in formal logic and to logicians without a mathematical background. 1961 edition. 112pp. 5³/₈ x 8¹/₂.
0-486-46249-8

INTRODUCTORY REAL ANALYSIS, A.N. Kolmogorov, S. V. Fomin. Translated by Richard A. Silverman. Self-contained, evenly paced introduction to real and functional analysis. Some 350 problems. 403pp. 5³/₈ x 8½.   0-486-61226-0

APPLIED ANALYSIS, Cornelius Lanczos. Classic work on analysis and design of finite processes for approximating solution of analytical problems. Algebraic equations, matrices, harmonic analysis, quadrature methods, much more. 559pp. 5³/₈ x 8½.   0-486-65656-X

AN INTRODUCTION TO ALGEBRAIC STRUCTURES, Joseph Landin. Superb self-contained text covers "abstract algebra": sets and numbers, theory of groups, theory of rings, much more. Numerous well-chosen examples, exercises. 247pp. 5³/₈ x 8½.   0-486-65940-2

QUALITATIVE THEORY OF DIFFERENTIAL EQUATIONS, V. V. Nemytskii and V.V. Stepanov. Classic graduate-level text by two prominent Soviet mathematicians covers classical differential equations as well as topological dynamics and ergodic theory. Bibliographies. 523pp. 5³/₈ x 8½.   0-486-65954-2

THEORY OF MATRICES, Sam Perlis. Outstanding text covering rank, nonsingularity and inverses in connection with the development of canonical matrices under the relation of equivalence, and without the intervention of determinants. Includes exercises. 237pp. 5³/₈ x 8½.   0-486-66810-X

INTRODUCTION TO ANALYSIS, Maxwell Rosenlicht. Unusually clear, accessible coverage of set theory, real number system, metric spaces, continuous functions, Riemann integration, multiple integrals, more. Wide range of problems. Undergraduate level. Bibliography. 254pp. 5³/₈ x 8½.   0-486-65038-3

MODERN NONLINEAR EQUATIONS, Thomas L. Saaty. Emphasizes practical solution of problems; covers seven types of equations. ". . . a welcome contribution to the existing literature. . . ."—*Math Reviews.* 490pp. 5³/₈ x 8½.   0-486-64232-1

MATRICES AND LINEAR ALGEBRA, Hans Schneider and George Phillip Barker. Basic textbook covers theory of matrices and its applications to systems of linear equations and related topics such as determinants, eigenvalues and differential equations. Numerous exercises. 432pp. 5³/₈ x 8½.   0-486-66014-1

LINEAR ALGEBRA, Georgi E. Shilov. Determinants, linear spaces, matrix algebras, similar topics. For advanced undergraduates, graduates. Silverman translation. 387pp. 5³/₈ x 8½.   0-486-63518-X

MATHEMATICAL METHODS OF GAME AND ECONOMIC THEORY: Revised Edition, Jean-Pierre Aubin. This text begins with optimization theory and convex analysis, followed by topics in game theory and mathematical economics, and concluding with an introduction to nonlinear analysis and control theory. 1982 edition. 656pp. 6⅛ x 9¼.   0-486-46265-X

SET THEORY AND LOGIC, Robert R. Stoll. Lucid introduction to unified theory of mathematical concepts. Set theory and logic seen as tools for conceptual understanding of real number system. 496pp. 5³/₈ x 8¼.   0-486-63829-4

Math—Decision Theory, Statistics, Probability

INTRODUCTION TO PROBABILITY, John E. Freund. Featured topics include permutations and factorials, probabilities and odds, frequency interpretation, mathematical expectation, decision-making, postulates of probability, rule of elimination, much more. Exercises with some solutions. Summary. 1973 edition. 247pp. 5⅜ x 8½.
0-486-67549-1

STATISTICAL AND INDUCTIVE PROBABILITIES, Hugues Leblanc. This treatment addresses a decades-old dispute among probability theorists, asserting that both statistical and inductive probabilities may be treated as sentence-theoretic measurements, and that the latter qualify as estimates of the former. 1962 edition. 160pp. 5⅜ x 8½.
0-486-44980-7

APPLIED MULTIVARIATE ANALYSIS: Using Bayesian and Frequentist Methods of Inference, Second Edition, S. James Press. This two-part treatment deals with foundations as well as models and applications. Topics include continuous multivariate distributions; regression and analysis of variance; factor analysis and latent structure analysis; and structuring multivariate populations. 1982 edition. 692pp. 5⅜ x 8½.
0-486-44236-5

LINEAR PROGRAMMING AND ECONOMIC ANALYSIS, Robert Dorfman, Paul A. Samuelson and Robert M. Solow. First comprehensive treatment of linear programming in standard economic analysis. Game theory, modern welfare economics, Leontief input-output, more. 525pp. 5⅜ x 8½.
0-486-65491-5

PROBABILITY: AN INTRODUCTION, Samuel Goldberg. Excellent basic text covers set theory, probability theory for finite sample spaces, binomial theorem, much more. 360 problems. Bibliographies. 322pp. 5⅜ x 8½.
0-486-65252-1

GAMES AND DECISIONS: INTRODUCTION AND CRITICAL SURVEY, R. Duncan Luce and Howard Raiffa. Superb nontechnical introduction to game theory, primarily applied to social sciences. Utility theory, zero-sum games, n-person games, decision-making, much more. Bibliography. 509pp. 5⅜ x 8½.
0-486-65943-7

INTRODUCTION TO THE THEORY OF GAMES, J. C. C. McKinsey. This comprehensive overview of the mathematical theory of games illustrates applications to situations involving conflicts of interest, including economic, social, political, and military contexts. Appropriate for advanced undergraduate and graduate courses; advanced calculus a prerequisite. 1952 ed. x+372pp. 5⅜ x 8½.
0-486-42811-7

FIFTY CHALLENGING PROBLEMS IN PROBABILITY WITH SOLUTIONS, Frederick Mosteller. Remarkable puzzlers, graded in difficulty, illustrate elementary and advanced aspects of probability. Detailed solutions. 88pp. 5⅜ x 8½.
0-486-65355-2

PROBABILITY THEORY: A CONCISE COURSE, Y. A. Rozanov. Highly readable, self-contained introduction covers combination of events, dependent events, Bernoulli trials, etc. 148pp. 5⅝ x 8¼.
0-486-63544-9

THE STATISTICAL ANALYSIS OF EXPERIMENTAL DATA, John Mandel. First half of book presents fundamental mathematical definitions, concepts and facts while remaining half deals with statistics primarily as an interpretive tool. Well-written text, numerous worked examples with step-by-step presentation. Includes 116 tables. 448pp. 5⅜ x 8½.
0-486-64666-1

Math—Geometry and Topology

ELEMENTARY CONCEPTS OF TOPOLOGY, Paul Alexandroff. Elegant, intuitive approach to topology from set-theoretic topology to Betti groups; how concepts of topology are useful in math and physics. 25 figures. 57pp. 5³/₈ x 8¹/₂.   0-486-60747-X

A LONG WAY FROM EUCLID, Constance Reid. Lively guide by a prominent historian focuses on the role of Euclid's Elements in subsequent mathematical developments. Elementary algebra and plane geometry are sole prerequisites. 80 drawings. 1963 edition. 304pp. 5³/₈ x 8¹/₂.   0-486-43613-6

EXPERIMENTS IN TOPOLOGY, Stephen Barr. Classic, lively explanation of one of the byways of mathematics. Klein bottles, Moebius strips, projective planes, map coloring, problem of the Koenigsberg bridges, much more, described with clarity and wit. 43 figures. 210pp. 5³/₈ x 8¹/₂.   0-486-25933-1

THE GEOMETRY OF RENÉ DESCARTES, René Descartes. The great work founded analytical geometry. Original French text, Descartes's own diagrams, together with definitive Smith-Latham translation. 244pp. 5³/₈ x 8¹/₂.   0-486-60068-8

EUCLIDEAN GEOMETRY AND TRANSFORMATIONS, Clayton W. Dodge. This introduction to Euclidean geometry emphasizes transformations, particularly isometries and similarities. Suitable for undergraduate courses, it includes numerous examples, many with detailed answers. 1972 ed. viii+296pp. 6¹/₈ x 9¹/₄.   0-486-43476-1

EXCURSIONS IN GEOMETRY, C. Stanley Ogilvy. A straightedge, compass, and a little thought are all that's needed to discover the intellectual excitement of geometry. Harmonic division and Apollonian circles, inversive geometry, hexlet, Golden Section, more. 132 illustrations. 192pp. 5³/₈ x 8¹/₂.   0-486-26530-7

THE THIRTEEN BOOKS OF EUCLID'S ELEMENTS, translated with introduction and commentary by Sir Thomas L. Heath. Definitive edition. Textual and linguistic notes, mathematical analysis. 2,500 years of critical commentary. Unabridged. 1,414pp. 5³/₈ x 8¹/₂. Three-vol. set.
            Vol. I: 0-486-60088-2   Vol. II: 0-486-60089-0   Vol. III: 0-486-60090-4

SPACE AND GEOMETRY: IN THE LIGHT OF PHYSIOLOGICAL, PSYCHOLOGICAL AND PHYSICAL INQUIRY, Ernst Mach. Three essays by an eminent philosopher and scientist explore the nature, origin, and development of our concepts of space, with a distinctness and precision suitable for undergraduate students and other readers. 1906 ed. vi+148pp. 5³/₈ x 8¹/₂.   0-486-43909-7

GEOMETRY OF COMPLEX NUMBERS, Hans Schwerdtfeger. Illuminating, widely praised book on analytic geometry of circles, the Moebius transformation, and two-dimensional non-Euclidean geometries. 200pp. 5⁵/₈ x 8¹/₄.   0-486-63830-8

DIFFERENTIAL GEOMETRY, Heinrich W. Guggenheimer. Local differential geometry as an application of advanced calculus and linear algebra. Curvature, transformation groups, surfaces, more. Exercises. 62 figures. 378pp. 5³/₈ x 8¹/₂.   0-486-63433-7

History of Math

THE WORKS OF ARCHIMEDES, Archimedes (T. L. Heath, ed.). Topics include the famous problems of the ratio of the areas of a cylinder and an inscribed sphere; the measurement of a circle; the properties of conoids, spheroids, and spirals; and the quadrature of the parabola. Informative introduction. clxxxvi+326pp. 5³⁄₈ x 8¹⁄₂.   0-486-42084-1

A SHORT ACCOUNT OF THE HISTORY OF MATHEMATICS, W. W. Rouse Ball. One of clearest, most authoritative surveys from the Egyptians and Phoenicians through 19th-century figures such as Grassman, Galois, Riemann. Fourth edition. 522pp. 5³⁄₈ x 8¹⁄₂.   0-486-20630-0

THE HISTORY OF THE CALCULUS AND ITS CONCEPTUAL DEVELOP-MENT, Carl B. Boyer. Origins in antiquity, medieval contributions, work of Newton, Leibniz, rigorous formulation. Treatment is verbal. 346pp. 5³⁄₈ x 8¹⁄₂.   0-486-60509-4

THE HISTORICAL ROOTS OF ELEMENTARY MATHEMATICS, Lucas N. H. Bunt, Phillip S. Jones, and Jack D. Bedient. Fundamental underpinnings of modern arithmetic, algebra, geometry and number systems derived from ancient civilizations. 320pp. 5³⁄₈ x 8¹⁄₂.   0-486-25563-8

THE HISTORY OF THE CALCULUS AND ITS CONCEPTUAL DEVELOP-MENT, Carl B. Boyer. Fluent description of the development of both the integral and differential calculus—its early beginnings in antiquity, medieval contributions, and a consideration of Newton and Leibniz. 368pp. 5³⁄₈ x 8¹⁄₂.   0-486-60509-4

GAMES, GODS & GAMBLING: A HISTORY OF PROBABILITY AND STATISTICAL IDEAS, F. N. David. Episodes from the lives of Galileo, Fermat, Pascal, and others illustrate this fascinating account of the roots of mathematics. Features thought-provoking references to classics, archaeology, biography, poetry. 1962 edition. 304pp. 5³⁄₈ x 8¹⁄₂. (Available in U.S. only.)   0-486-40023-9

OF MEN AND NUMBERS: THE STORY OF THE GREAT MATHEMATICIANS, Jane Muir. Fascinating accounts of the lives and accomplishments of history's greatest mathematical minds—Pythagoras, Descartes, Euler, Pascal, Cantor, many more. Anecdotal, illuminating. 30 diagrams. Bibliography. 256pp. 5³⁄₈ x 8¹⁄₂.   0-486-28973-7

HISTORY OF MATHEMATICS, David E. Smith. Nontechnical survey from ancient Greece and Orient to late 19th century; evolution of arithmetic, geometry, trigonometry, calculating devices, algebra, the calculus. 362 illustrations. 1,355pp. 5³⁄₈ x 8¹⁄₂. Two-vol. set.   Vol. I: 0-486-20429-4 Vol. II: 0-486-20430-8

A CONCISE HISTORY OF MATHEMATICS, Dirk J. Struik. The best brief history of mathematics. Stresses origins and covers every major figure from ancient Near East to 19th century. 41 illustrations. 195pp. 5³⁄₈ x 8¹⁄₂.   0-486-60255-9

CATALOG OF DOVER BOOKS

A TREATISE ON ELECTRICITY AND MAGNETISM, James Clerk Maxwell. Important foundation work of modern physics. Brings to final form Maxwell's theory of electromagnetism and rigorously derives his general equations of field theory. 1,084pp. 5⅜ x 8½. Two-vol. set. Vol. I: 0-486-60636-8 Vol. II: 0-486-60637-6

MATHEMATICS FOR PHYSICISTS, Philippe Dennery and Andre Krzywicki. Superb text provides math needed to understand today's more advanced topics in physics and engineering. Theory of functions of a complex variable, linear vector spaces, much more. Problems. 1967 edition. 400pp. 6½ x 9¼.   0-486-69193-4

INTRODUCTION TO QUANTUM MECHANICS WITH APPLICATIONS TO CHEMISTRY, Linus Pauling & E. Bright Wilson, Jr. Classic undergraduate text by Nobel Prize winner applies quantum mechanics to chemical and physical problems. Numerous tables and figures enhance the text. Chapter bibliographies. Appendices. Index. 468pp. 5⅜ x 8½.   0-486-64871-0

METHODS OF THERMODYNAMICS, Howard Reiss. Outstanding text focuses on physical technique of thermodynamics, typical problem areas of understanding, and significance and use of thermodynamic potential. 1965 edition. 238pp. 5⅜ x 8½.
                                                                                        0-486-69445-3

THE ELECTROMAGNETIC FIELD, Albert Shadowitz. Comprehensive under- graduate text covers basics of electric and magnetic fields, builds up to electromagnetic theory. Also related topics, including relativity. Over 900 problems. 768pp. 5⅝ x 8¼.
                                                                                        0-486-65660-8

GREAT EXPERIMENTS IN PHYSICS: FIRSTHAND ACCOUNTS FROM GALILEO TO EINSTEIN, Morris H. Shamos (ed.). 25 crucial discoveries: Newton's laws of motion, Chadwick's study of the neutron, Hertz on electromagnetic waves, more. Original accounts clearly annotated. 370pp. 5⅜ x 8½.   0-486-25346-5

EINSTEIN'S LEGACY, Julian Schwinger. A Nobel Laureate relates fascinating story of Einstein and development of relativity theory in well-illustrated, nontechnical volume. Subjects include meaning of time, paradoxes of space travel, gravity and its effect on light, non-Euclidean geometry and curving of space-time, impact of radio astronomy and space-age discoveries, and more. 189 b/w illustrations. xiv+250pp. 8⅜ x 9¼.   0-486-41974-6

THE VARIATIONAL PRINCIPLES OF MECHANICS, Cornelius Lanczos. Philosophic, less formalistic approach to analytical mechanics offers model of clear, scholarly exposition at graduate level with coverage of basics, calculus of variations, principle of virtual work, equations of motion, more. 418pp. 5⅜ x 8½.   0-486-65067-7

CPSIA information can be obtained
at www.ICGtesting.com
Printed in the USA
BVHW091059131118
532891BV00011B/444/P